SPATIO-TEMPORAL CHARACTERISATION OF DROUGHT:

DATA ANALYTICS, MODELLING, TRACKING, IMPACT AND PREDICTION

Vitali Díaz Mercado

SPATIO-TEMPORAL CHARACTERISATION OF DROUGHT:
DATA ANALYTICS, MODELLING, TRACKING, IMPACT AND PREDICTION

DISSERTATION

Submitted in fulfilment of the requirement of
the Board for Doctorates of Delft University of Technology
and of
the Academic Board of IHE Delft Institute for Water Education
for the Degree of DOCTOR
to be defended in public on
Wednesday, 24 November 2021 at 10:00 hours
in Delft, the Netherlands

by

Vitali DÍAZ MERCADO

Master of Science in Water Science
Universidad Autónoma del Estado de México

born in Toluca, México

This research was conducted under the auspices of the Graduate School for Socio-Economic and Natural Sciences of the Environment (SENSE)

Published by:
CRC Press/Balkema
enquiries@taylorandfrancis.com
www.crcpress.com – www.taylorandfrancis.com

ISBN 978-1-032-24650-5

To my loves Febe, Simcha and Nathan

ACKNOWLEDGMENTS

Behind the development of this research are a good number of people who I love and appreciate. First, I want to express my gratitude to my God and Lord Jesus Christ. My wife, close friends and relatives, witnessed what it cost to start this adventure, which would not have been possible without my God, to Him be the glory.

This dissertation is dedicated to you, my beloved wife Febe. Thank you for your constant support, dedication, long and deep talks and sharp observations. Thank you for this time in this country. Some time ago, we began this adventure with many unknowns but with great assurance that we were supported. I love you Febe, you and Simcha are my reason to move on.

Gracias mamá, gracias papá. Este tiempo lejos de México nos ha enseñado muchas cosas. Gracias por sus oraciones constantes y su ayuda. He aprendido mucho de ustedes lo cual me ha ayudado en este tiempo viviendo aquí en este país. Amados suegros les doy las gracias, sus oraciones y ayuda incondicional ha sido un gran soporte en este tiempo. Gracias por su apoyo en todos los trámites necesarios en México a lo largo de este periodo, siempre estuvieron cerca. Nahin, gracias por tu apoyo y por cuidar a mis papás, te amo y extraño carnalito. Gracias Marce y Raquel por su apoyo, nos han ayudado mucho, los amo.

My special gratitude goes to you, Professor Dimitri. Thank you for your supervision, your always so challenging reflections, your consistently accurate and sharp comments. I am also very grateful for your understanding and support in my absences from work during my son's birth and first months. You have always seen your students as people in professional and personal development. Thank you for all your end-of-year letters and all the details that you always had for me.

To you, Gerald, ¡gracias amigo! thank you very much for accepting me as your PhD student. You have helped me a lot, not only professionally but also personally. Thank you for your always pointed comments, for all those long talks to develop this research. You have a lot of great ideas, thanks for sharing some of them. Throughout my PhD you have invited me to carry out teaching, supervision, research activities; thank you for that detail. I have learned a lot. These activities would not have been possible without your intervention and guidance. Gracias.

I was fortunate to meet very wonderful and talented people who contributed directly and indirectly to the improvement and execution of this research. Thank you Henny for your collaboration, your comments helped a lot to improve and clarify this research. Thank you for agreeing to be a speaker at the IHE PhD symposium, where I was a co-organiser. Thanks for translating the abstract. Thank you for your good wishes that you always sent me through your emails. Thank you very much Henny. Thank you very much Emmanouil for your observations, suggestions, your time in reviews. You were always attentive not only to the academics but also to personal matters, thank you. Thank you Shreedhar for your comments and suggestions to develop this research. Thanks Leonardo for your advice and recommendations, especially in the stage where my son was born. Ilyas, thank you very much for your support and advice. You were always very attentive to my well-being and family. Angeles, thank you very much for all your help, for the meetings you organised with the Mexican community. I think that helped many of us to have a much better time in Delft. Thank you also for all your attentions. Carlos, thank you very much for your help, you were always so attentive and kind. You were our first point of contact at IHE, and from where this whole adventure began, thank you. Roos, I did not have the opportunity to tell you how much you helped me, your talks at IHE made this place gezellig.

My time at IHE Delft was more enjoyable thanks to the company of many colleagues who, with their talks, suggestions and jokes, made the days pleasant. To all those who anonymously collaborated with the gift when my son was born, I thank you. Thanks to everyone who always had a word for academic and personal matters. There are several names and I would not like to forget any, apologies in advance. Thanks to Jonatan, José Luis, Mario, Juan Carlos, Miguel, Gonzalo, Francisco, Neiler, Natalia, Sara, Alex, Meseret, Vo, Mohaned, Carlos, Pan, Ana María, Duoc, Shahnoor, Thaine, Ha, Adriano, Mauricio, Alex, Yared, Musaed, Taha, Akosua, Aline, Maria and Alessandro. Special thanks to you Victor Paca, for your pleasant talks about Mexico, its history, its music, about life, for the time we philosophised while the scripts ran, thank you very much Victor.

I am also grateful to all those I had the fortune to supervise and teach. Through your questions, comments, and suggestions, I improved many aspects of my concepts and myself. Many ideas on drought were much more precise after the supervision\teaching exercise. Thanks to Ahmed Osman, Yousra Omer, Manh-Hung. Ricardo, thank you very much for your stay in Delft, you do not know how much you helped me. In my last stage, I met you Ali, thank you for your professionalism, suggestions, patience; all the best to you dear Ali.

Undoubtedly, many aspects of my stay at IHE Delft were well-accomplished thanks to the collaboration and professionalism of its lecturers and staff. My sincere thanks to Jolanda, Maria Laura, Anique, Floor, Ellen, Conny, Cristina, Gaetano, Michael, William, Ioana, Schalk Jan, Biswa, Andreja, Gerda, Mita, Willem, Eldon, Dominique, Ruth, Ewoud, and Eric. Thanks to Paula Derkse for her guidance and comments to improve this document for printing.

I thank IHE Delft for being so inclusive. The Biblical studies that we carried out within its facilities by the student community and some staff were beneficial, both in difficult moments and in moments of sharing some joy. Thanks to all the students who shared their testimonies, your energy and dedication touched my heart, being in another country is not always easy, but you encouraged me with your examples. Thanks Miranda for all the meetings you organised and where I participated.

Many thanks to the ICF Delft community and friends in Delft. All of you helped us in our day to day in this country. Thanks to Pastor Niek Tramper for his wise teachings and constant visits to see that we all were well. We felt very protected under your care. Thank you David and Fera for your support and help when Febe was pregnant and during our stay in Delft. Thanks to Henk Polinder for all your attentions. Thanks to Pastor Hans-Jan, your teachings always gave me strength. Thank you very much to all of you guys, Arjan, Nicolas, Vincent, Darli, Peter, Haiyan, Mike and all who shared their time throughout the Bible studies with dinner and coffee. Thank you Conchi Moreno and husband, you helped us a lot. To you also Leila and Jose for your always so pleasant conversations and all your help. Thanks Adriana and Diego for your enjoyable company and talks.

For the execution of this PhD research, I was supported by the National Council of Science and Technology (Conacyt)-Alianza Fiidem doctoral scholarship (217776/382365). I am grateful to the Secretariat of Public Education (SEP) for the "complementary scholarship" (BC-4037) granted in the first year of my PhD. I thank the Science and Technology Council of Mexico State (Comecyt) for the grant 14BAE098 for accommodation expenses and the "apoyo a estudiates mexiquenses" grant (no. 21EI1731, 21EI1733 and 21EI1734). In the last stage of my PhD, I was supported by the "Uncertainty-aware intervention design for Mediterranean aquifer recharge" project, granted by the Prince Albert II of Monaco Foundation (No. 2579). I also thank my former MSc supervisor and friend Dr. Khalidou M. Ba for the financial support provided for the presentation of part of this research at the International Congress of Hydroinfomatics in Palermo, Italy (2018), through the project "Hydrology in Mauritania: Runoff modeling with satellite precipitation" (no. 4192/2016E).

I also want to thank the care and guidance that many colleagues are kind enough to offer through social media. Their advice, best practices, guidelines, recommendations, jokes, shortcuts, tips, recipes, photos, their sincerity in failures, how to deal with paper rejection, and their joy in the achievements, help me in many ways during my PhD journey. Especially thanks to @AcademicChatter, @ithinkwellHugh, @WriteThatPhD, Raul Pacheco-Vega's blog (http://www.raulpacheco.org/resources/), @dsquintana, @KinarNicholas, @AlbertoCairo, @GeostatsGuy, @mathladyhazel, and more.

Finalmente quiero concluir con unas líneas dedicadas a todos mis seres queridos en México, nuestros caminos se fueron por otros rumbos pero no nuestros corazones. Me vine un día sin saber que era el último para ver a muchos de ustedes, hoy ya no están físicamente pero laten profundo en mi corazón. Estos dos años han sido duros, mi abrazo incondicional a ustedes mis hermanos, y para los que han partido, Dios los tenga en su gloria.

Vitali, Delft, 2021

SUMMARY

Defining drought is not straightforward. Unlike with other natural phenomena, defining drought depends on the area or sector being affected. From the perspective of human and social needs, drought has been studied in its relation to water resource sectors, including agricultural, economic and environmental. These various foci link the different sectors to a specific water cycle process. Such processes help describe meteorological drought, for example, which refers to a water deficit caused by anomalous precipitation and temperatures. This definition is used by the general public and very often by water agencies. There is a clear connection between lack of precipitation, surface runoff deficits and river flow, lake and reservoir levels. Thus, calculations of surface runoff are at the core of hydrological drought studies. Agricultural drought, as the name implies, relates mainly to the lack of soil moisture that affects crops.

While the various drought types and their studies have resulted in different definitions, a good starting point is to refer to drought as a water deficit caused by anomalous precipitation and temperature. When extended over a region, this deficit can trigger a lack of soil moisture, runoff and groundwater. Importantly, intrinsic drought characteristics, like the space–time component of such a definition, are implicit. Characteristics of drought, such as duration and spatial extent, can be further extended to develop ideas regarding such concepts as drought similarity, morphology and dynamics.

Drought indicators are the most common tools used to calculate and monitor drought. Based on the principle of an above-normal deficit (i.e. an anomaly), these indicators use mathematical formulas to transform hydro-meteorological variables into statistical values. These values are then related to the expected range of a normal condition and used to identify anomalies. The hydrological process is analysed, and the applicable drought indicator used according to the type of drought to be assessed. Due to data availability issues, some types of drought can only be monitored using a proxy hydrological variable. Such proxies are evaluated according to ranges of anomalies that mirror the effects of the target variable. This allows different types of droughts to be monitored, even in the absence of direct data.

Along with the development of indicators and methods for drought monitoring, several contributions have been made to improve drought characterisation. Enhancing drought characterisation refers not only to improving estimations of the phenomenon's intensity, duration and spatial extent but also to increasing our knowledge and understanding of how droughts develop and change over time. The former is important for operational purposes, and the latter for scientific research.

In the last decades, studies of drought have increased in light of new data availability and advances in spatio-temporal analysis. However, the following gaps still need to be filled:

1) methods to characterise drought that explicitly consider its spatio-temporal features, such as area and pathway; 2) methods to monitor and predict drought that include the above-mentioned characteristics and 3) approaches for visualising and analysing drought characteristics to facilitate the interpretation of drought variations.

This research aims to explore, analyse and propose improvements to the spatio-temporal characterisation of drought. The improved characterisation, monitoring and visualisation of drought are expected to provide new perspectives and information towards better prediction. This research proposes the following objectives:

O1. Improve the methodology for characterising drought in space and time based on the phenomenon's spatial features, such as spatial extent and location.

O2. Develop a visual approach to analysing variations of spatio-temporal drought characteristics.

O3. Develop a methodology for monitoring the spatial extent of drought (i.e. drought tracking).

O4. Explore the applicability of using machine learning techniques to predict crop-yield responses to drought based on spatio-temporal drought characteristics.

In this dissertation, drought is conceptualised as a phenomenon with a spatial extent, onset and end in space and time, as well as a spatial path composed of the union of successive tracks. Machine learning (ML) techniques and process-based approaches are used to build the research methodology.

The present methodology was designed to achieve the proposed objectives. First, a literature review served as an overview of the existing methods and concepts regarding drought characterisation. The most relevant of those methods were selected to pursue the research objectives. For the first objective (O1), an approach was built to calculate and characterise drought. Different drought indicators were used to calculate meteorological, hydrological and agricultural droughts. For the second objective (O2), radial and polar charts were developed to analyse drought variations. Each visual element of the graphs encoded different drought characteristics, such as intensity and spatial extent. For the third objective (O3), a drought tracking method was constructed to calculate drought trajectories in space. The most extended droughts were calculated at a country scale and analysed using this new method. For the fourth objective, (O4) ML models were built to predict seasonal crop yields using drought areas as the main input.

The first outcome compared different types of droughts using two approaches: area-aggregated drought indicators and drought areas. Comparisons were carried out at the basin scale, at which meteorological, hydrological and agricultural droughts were analysed. The results showed little difference among the time series of the area-aggregated drought indicators. However, the drought areas showed variations among the types of drought analysed. The drought area approach clearly detected a lag of time between the meteorological and hydrological droughts. Comparisons across drought areas also revealed different seasonal drought behaviours, which were not easily detectable through aggregate time series.

The second outcome was the construction of the season crop-yield model. In this second application, an ML framework for seasonal crop-yield prediction was built using drought areas. The drought areas were calculated from a database produced by a well-known global drought monitor and used to predict crop yields in three regions. The results showed that drought area is a suitable variable because its size is a good indication of drought magnitude.

For the third outcome, three visual approaches were developed based on radial and polar charts for drought analysis. These visual tools helped to assess drought variations in the face of challenges specific to this type of analysis. The challenges included analysing more than two drought characteristics in a large data period, or identifying patterns, such as seasonality, that are not straightforward in line or area graphs. The results indicated that the proposed charts help to identify drought intensity and trends. These graphs allowed difficult patterns, such as seasonality, to be spotted more easily.

The fourth outcome was the methodology for building the spatial path of a drought (i.e. the union of successive tracks). This methodology was applied to a case study of India, analysing the largest events from 1901 to 2013. The occurrence of the calculated droughts was corroborated with documented information from the region. For each drought event, the onset and final location, direction and spatial path were calculated. The data generated by the tracking methodology were then used to characterise drought dynamics. The results from India showed that consecutive areas in time overlapped considerably, suggesting that the spatial extent of drought remains in the same region after reaching a considerable size. The presence of large drought areas in the same region over time may explain the severity of such droughts.

Finally, a scope was formulated for integrating ML and the spatio-temporal analysis of droughts. The proposed scope opens a new area of potential for drought prediction (i.e. predicting spatial drought tracks and areas). It is expected that the drought tracking and prediction method, when fully completed, will help populations cope with drought and its severe impacts.

SAMENVATTING

Droogte treft meestal uitgestrekte gebieden en beperkt zich vrijwel nooit tot landsgrenzen. Droogte kan overal op de wereld voorkomen met ernstige gevolgen voor watervoorraden (oppervlaktewater en grondwater) en sociaaleconomische activiteiten. Het is alom bekend dat een betere karakterisering van droogte zal leiden tot een succesvollere ontwikkeling en implementatie van beleidsmaatregelen voor het verminderen van de gevolgen van droogte. Een verbetering van de karakterisering impliceert het vaststellen van de intensiteit, duur en uitgestrektheid van de droogte. Ofschoon, de karakterisering is gestimuleerd door het beschikbaar komen van nieuwe gegevens en vooruitgang in ruimtelijk-temporele analyse, de volgende leemten in kennis bestaan nog steeds: (1) ontbreken van methoden, die expliciet de ruimtelijk-temporele eigenschappen van droogte karakteriseren, zoals de oppervlakte in droogte en de verplaatsing (traject) van de droogte, (2) ontbreken van methoden, die deze eigenschappen monitoren en voorspellen, en (3) verbetering van systemen, die eigenschappen van droogte visualiseren/analyseren t.b.v. de interpretatie van ruimtelijk-temporele patronen. Dit onderzoek draagt bij aan kennisontwikkeling door de ontwikkeling en toepassing van visualisatie methodieken om de ruimtelijk-temporele karakterisering van droogte (STAND) uit te kunnen voeren en helpt de voordelen te evalueren wanneer de methodieken worden gebruikt voor voorspelling. Een uitgebreider concept voor droogte wordt geïntroduceerd, namelijk als een gebeurtenis (onderwerp) met een oppervlakte, een begin en een einde in ruimte en de tijd en een verplaatsing die bestaat uit samengestelde, opeenvolgende trajecten. De ontwikkelde methodieken maken gebruik van zelflerende technieken en proces-georiënteerde benaderingen. De methodieken worden geïllustreerd met de volgende voorbeelden: (1) voordelen van het gebruik van gebieden in droogte versus ruimtelijk-gemiddelde droogte indices, (2) een veelbelovende, zelflerende techniek waarbij tijdreeksen van gebieden in droogte worden gebruikt om de gevolgen van droogte op gewasopbrengsten te voorspellen, (3) analyse van ruimtelijk-temporele droogtepatronen door visualisatie methodieken, en (4) analyse van de dynamiek in ruimtelijke patronen van de meest ernstige droogte op de nationale schaal (India), inclusief verplaatsingen. Tenslotte, wordt aan de hand van dit proefschrift een perspectief geschetst voor de integratie van zelflerende technieken en ruimtelijk-temporele droogte resultaten om nieuwe onderzoeksgebieden op het terrein van voorspelling van droogte verplaatsingen verder te ontwikkelen.

CONTENTS

1

INTRODUCTION

1.1 BACKGROUND

1.1.1 Drought

Drought is a regional phenomenon that can occur anywhere in the world and often has severe consequences in terms of water resources and socioeconomic activity (Below et al., 2007; Markonis et al., 2013; Mishra and Singh, 2010; Sheffield and Wood, 2011; Tallaksen and Van Lanen, 2004; Van Lanen et al., 2013; Wilhite, 2000). It has been widely recognised that improving drought analyses would allow the development and implementation of more successful national policies for mitigating drought impacts (World Meteorological Organization (WMO), 2006). The WMO points out that to reduce the negative impacts of drought, technologies and methods must be developed to improve drought's characterisation (i.e. enhance the calculation of its duration, magnitude, spatial extent, onset and end, among others).

There is no unique definition of drought. However, there is an agreement among explanations that it is an anomaly in water availability, most often caused by changes in precipitation and temperature that lead to a lack of soil moisture, runoff and groundwater (Mishra and Singh, 2010; Van Loon, 2015). This lack of water availability is expressed and often analysed through a drought indicator. Drought indicators transform the hydro-meteorological variable into a value that relates to a measurable standard statistic, allowing drought to be identified as an anomaly (Mshra and Singh, 2011).

These drought indicators are commonly used to perform drought characterisation. The procedure of characterisation starts with drought calculations. First, the hydrometeorological variable is selected, and then transformed into the drought indicator. For each of the calculated droughts, an onset and an end are computed; these allow drought duration to be calculated. Drought-indicator magnitudes also help to estimate the severity of each calculated drought (Mishra and Singh, 2011).

Currently, in many countries regional drought monitoring is often conducted using drought-monitoring systems, which are fed with hydro-meteorological variables to compute the drought indicators and estimate drought characteristics (Hao et al., 2017). The spatial condition of drought, including its extent, is monitored with the help of time snapshots, which provide qualitative information on the spatial behaviour of the phenomenon.

The available drought monitors deliver information about spatial extent (i.e. snapshots); however, there still lacks a consistent procedure for assessing the variations of spatio-temporal dynamics (Hao et al., 2017). 'Drought dynamics' refers to the way in which the spatial distribution of a drought changes over time. Developing new technologies to increase understandings of how droughts develop in space and time may help to acquire knowledge on their drivers and processes; such information would help improve drought monitoring and prediction. Thus, this research investigates spatio-temporal drought dynamics.

1.1.2 Drought study

The study of drought consists of different components, such as drought characterisation, employing the methods to calculate drought features such as duration, magnitude and spatial extent. To provide context, the following paragraphs briefly describe the relationship between drought characterisation and other forms of drought study. Notably, some concepts related to drought characterisation have significant overlap.

Drought study components

Figure 1.1 presents a scheme showing the main components of drought study. The classifications are based on the respective objective that each component seeks within drought research. The overlapping areas in Figure 1.1 indicate the methods and models shared by different components. Drought study comprises drought indicators, characterisation, monitoring, prediction, analysis, visualisation and impact assessment.

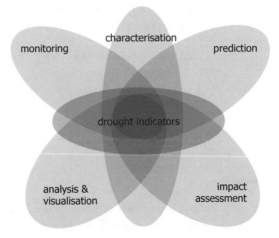

Figure 1.1 Different components of drought study.

In each of the components shown in Table 1.1, two types of models can be used: process-based and machine learning (ML) models. Most drought indicators focus on the identification and calculation of water anomalies in specific parts of the hydrological cycle. In drought monitoring and forecasting, process-based models are used to determine various hydro-meteorological variables needed to calculate water anomalies. When analysing the impact of drought on such areas as crops, models based on growth stages are used (i.e. crop phenology, which serves to monitor and estimate drought effects).

Table 1.1 Objective of drought study components

Component	Objective
Drought indicators	Calculation, development, improvement and testing of drought indicators.
Characterisation	Calculation of drought characteristics (e.g. duration, severity, spatial extent).
	Development of methods to describe drought development, changes and patterns.
Monitoring	Systematic review, observation, evaluation and reporting of drought development and/or effects (impacts) over a period of time.
Prediction	Systematic calculation, evaluation, model tuning and reporting of drought development and/or effects (impacts) over a forecasted time window.
Analysis and visualisation	Detailed examination of drought indicators, characteristics, development and effects (impacts), as well as their relationships.
Impact assessment	Calculation, evaluation and analysis of drought effects on environmental, economic and social activities.

1.1.3 Machine learning

When analysing drought, statistical and ML techniques are mainly used to predict (forecast) drought indicators and drought impacts. In characterising drought dynamics, literature reports a small number of examples of the use of ML for such purpose. ML models can be a useful complement to the physically-based (process) models, because they require less knowledge about the underlying complex physical processes, provided there is enough data to calibrate (train) them. (Learning in this context refers to the ability of a model to predict or classify information based on historical data.)

Different types of techniques can be identified in ML, such as supervised and unsupervised learning (Figure 1.2). In supervised learning, the model setup follows the training-calibration-validation procedure. Supervised learning techniques include long-known linear regression, as well as various types of neural networks, support vector machines, random forests, etc. In unsupervised learning, no training step is required because the goal is to find common elements among the data that help in their grouping (labelling). Clustering and principal components analysis are examples of unsupervised learning techniques (Talabis et al., 2015). This research applies both types of ML techniques, most notably clustering, polynomial regression and neural networks.

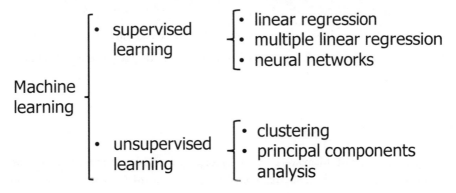

Figure 1.2 Types of ML techniques and examples. Figure shows a small list of techniques from the wide variety that exist.

The methodology for this research draws on both process-based and ML models. Due to the complexity of the processes involved in calculating and characterising drought, ML techniques are seen as an important element of the methodology. The research methodology does not follow a purely ML approach, however; the data analysis and subsequent construction and application of ML models are integrated with process-based models.

1.2 PROBLEM STATEMENT

Currently, drought monitoring and prediction are not fully aided by the spatio-temporal analytical techniques. This demonstrates an opportunity to improve the way drought dynamics is characterised to explicitly consider the spatio-temporal features, such as spatial extent (spatial boundaries), location and spatial pathway. Due to the complexity of the phenomenon's variation, approaches for visualising and analysing drought must also be improved to allow spatio-temporal drought patterns to be interpreted effectively.

1.3 RESEARCH HYPOTHESES

This research proposes the following key hypotheses:

1) The spatio-temporal characteristics of drought can be calculated effectively using ML techniques.

2) Drought tracking can be improved by considering its spatio-temporal characteristics, such as spatial extent, with the appropriate visualisation techniques.

1.4 RESEARCH OBJECTIVES

The main objective of this research is to explore, analyse and propose improvements to the characterisation of spatio-temporal drought dynamics. The characterisation, monitoring and visualisation of drought are expected to provide new perspectives and information towards improved prediction.

The research proposes the following objectives:

1) Improve the methodology for characterising drought in space and in time based on the phenomenon's spatial features, such as spatial extent and location.

2) Develop a visual approach to analysing the variation of spatio-temporal drought characteristics.

3) Develop a methodology for monitoring the spatial extent of drought, i.e. drought tracking.

4) Explore the applicability of using ML techniques to predict crop-yield responses to drought based on spatio-temporal drought characteristics.

1.5 DISSERTATION STRUCTURE

This dissertation consists of 10 chapters, as outlined in Figure 1.3. The arrows indicate how the chapters interlink.

Chapter 1: Introduction. This chapter presents the background, problem statement, objectives and hypothesis of the research.

Chapter 2: Literature review. Concepts regarding the components of drought study are described, as are the ML techniques used to build the research methodology.

Chapter 3: Methodological framework. This chapter presents the general framework of this research.

Chapter 4: Case studies and data. The cases studies and data selected for this research are described.

Chapter 5: Spatio-temporal drought characterisation. The Spatio-Temporal ANalysis of Drought (STAND) methodology for drought characterisation is introduced.

Chapter 6: Comparison of drought indicators. This chapter explains the methodology's application to analyse how the spatial extent of drought changes across the hydrological cycle, from precipitation to runoff. It introduces the Standardized Evaporation Deficit Index (SEDI) developed for drought monitoring. The benefits of considering drought areas over area-aggregated drought indices are presented.

Chapter 7: Machine-learning approach to crop yield prediction. An approach of using drought areas as inputs to predict crop yield is presented.

Chapter 8: Visual approaches to drought analysis. This chapter introduces approaches based on radial charts for visualising and analysing drought variation.

Chapter 9: Spatial drought tracking development. This chapter proposes an approach for characterising spatio-temporal drought dynamics and calculating the spatial paths of drought.

Chapter 10: Conclusions and recommendations. The main conclusions of the research are presented, and recommendations made for future work on the subject.

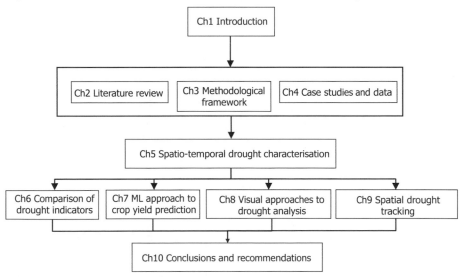

Figure 1.3 Structure of PhD dissertation.

2
LITERATURE REVIEW

2.1 DROUGHT

Drought is a natural phenomenon whose impacts generate many economic and human losses (Below et al., 2007; Sheffield and Wood, 2011; Tallaksen and van Lanen, 2004). A number of disciplines offer varying definitions of drought (Bachmair et al., 2016; Mishra and Singh, 2010), but the general consensus is that it is an anomaly originating in precipitation and temperature, whose further effects are observed in other components of the hydrological cycle (e.g. soil moisture, runoff, effect on human activities) (Tallaksen and van Lanen, 2004; van Loon, 2015). These definitions do not, however, explicitly elaborate on the spatial extent and duration of drought. Many studies of global models that explore drought require arbitrary steps to define the size (threshold) that can be used to consider an event as extreme or normal. In this sense, one drought event in a large region can be generalised in a particular time, even if the region only suffers drought in a small part of its area.

Drought indicators or indices (DIs) are typically used to estimate the phenomenon's magnitude and duration. A DI is a mathematical formulation that quantifies the water anomaly in a hydrometeorological variable (Mishra and Singh, 2010). When the DI is computed over a region in a spatial way, the spatial extent of the drought is estimated using a spatio-temporal method (e.g. Corzo Perez et al., 2011; Hannaford et al., 2011; Herrera-Estrada et al., 2017; Hisdal and Tallaksen, 2003; Lloyd-Hughes, 2012; Peters et al., 2006; Sheffield et al., 2009; Tallaksen et al., 2009; Tallaksen and Stahl, 2014; Van Huijgevoort et al., 2013; Vicente-Serrano, 2006; Zaidman et al., 2001). Per current practices, monitoring of drought magnitude, duration and area over a region is performed via drought monitoring systems. These systems are fed with hydro-meteorological variables, and the outcomes generated are known as drought-monitoring products. Drought indicator databases are part of this information. These drought-related data allow more detailed analyses to be performed on the spatio-temporal development of drought.

Various types of drought indicators have been proposed for identification of different types of droughts (Mishra and Singh, 2011; Wanders et al., 2010). Drought indicators can be grouped into two sets: those that use procedures for standardisation of water anomalies, which are expressed as drought indices; and those where the hydrometeorological variable is directly evaluated via a threshold, such as a given percentile, and are not indices in the strict sense. Extensive catalogues of drought indicators are presented in Wanders et al. (2010) and WMO and Global Water Partnership (GWP) (2016).

Variables that allow the water content in soil to be evaluated directly (e.g. soil moisture) are generally recommended when assessing agricultural drought (Agutu et al., 2017; Modanesi et al., 2020; WMO, 2011). Another variable, used to indirectly infer vegetation condition, is the Normalized Difference Vegetation Index (NDVI). This index is based on remote sensing data (Kogan, 1995). However, the NDVI can be influenced by factors such as topography, cloud cover and land cover (Kogan, 1995). The Vegetation Condition Index (VCI), based on the NDVI, was designed to overcome these limitations. Other remote sensing formulations for agricultural drought assessment include the Temperature Condition Index (TCI) (Kogan, 1995) and methods based on or inspired by the NDVI, VCI and TCI. Another drought indicator used for agricultural assessments is the Palmer Drought Severity Index (PDSI) proposed by Palmer (1965). In PDSI calculation, relative soil moisture (SM) is computed first by modelling the water-budget system. Soil characteristics, precipitation and evapotranspiration data are considered for its calculation. Next, SM is standardised via a procedure suggested by Palmer (1965), in which the long-term average components of the water-budget system are considered. This drought indicator has two key limitations. First, PDSI results are hardly comparable anywhere in the world because Palmer (1965) determined empirical constants using sample data from select parts of the United States to model the water-budget system. Second, snow melting is not considered when calculating PDSI, which restricts its use in some regions of the globe. To overcome these drawbacks, some improvements have been done that include a procedure to calibrate the water-budget system (Wells et al., 2004) and the integration of a snowmelt model (Van Der Schrier et al., 2013). These additions mean PDSI values can be used to compare drought variations across different regions. However, there are few applications of its use.

Yet another approach often applied to calculate agricultural drought indicators is to use meteorological drought indicators as proxies of the former ones (WMO, 2012). Meteorological variables, such as precipitation, are accumulated over an aggregation period (1, 3, 6, 9 or 12 months) to infer what could happen in the terrestrial hydrological cycle. When using meteorological indicators as proxies, various aggregation periods are typically tested to determine the optimum length for agricultural drought assessment.

2.2 DROUGHT INDICATORS

Drought is identified and quantified through the application of drought indicators. In this dissertation, the term 'drought indicator' refers both to drought indices and other procedures like percentile-based threshold levels. Drought indicators are mathematical formulations that assign a number to the magnitude of the water deficit. Table 2.1 presents the main types of droughts and the hydrometeorological variables that are often used to monitor them.

Meteorological drought indicators mainly use meteorological variables, such as precipitation and temperature, for their calculation. More information is often needed to compute agricultural and hydrological indicators. This information can come from observations or modelling. Meteorological indicators are also used to monitor hydrological and agricultural drought. To this end, the input when calculating a meteorological drought indicator is the meteorological variable with the time aggregation period equal to or longer than a month. The WMO (2012) presents different aggregation periods that can be used to monitor meteorological, hydrological and agricultural drought in terms of precipitation (Table 2.1, third column).

Several studies have described and compared drought indicators and their performance in detecting the different types of drought. Readers can consult the reviews of drought indicators by Keyantash and Dracup (2002), Wanders et al. (2010), Maskey and Trambauer (2014) and Bachmair et al. (2016). The following paragraphs provide a brief review of some of these studies.

Table 2.1 Types of drought and the hydrometeorological variables used to monitor them. P, E, R and SM stand for precipitation, evaporation, runoff and soil moisture, respectively. The third column applies to precipitation.

Type of drought	Hydrometeorological variable used in the monitoring	Time aggregation period of P in months
Meteorological	P, P-E, E	1,3
Agricultural	SM	1 to 6
Hydrological	R	6, 9+

Keyantash and Dracup (2002) compared seven meteorological, five hydrological and six agricultural drought indicators using time series of precipitation, streamflow and soil moisture, respectively. The objective of their study was to find the most outstanding indicators for each type of drought. To evaluate the performance of drought indicators, they followed a weighted set of six evaluation criteria. Evaluations were performed using visual comparisons. They found that the best drought indicators for meteorological, hydrological and agricultural drought were based on precipitation, streamflow, and soil moisture, respectively.

Wanders et al. (2010) reviewed drought indicators that can be used for global-scale applications. They used time-series simulations of runoff and soil moisture from virtual catchments selected across the globe. The virtual catchments refer to half-degree cells with similar hydrological structures. Simulations were computed using a conceptual hydrological model forced with precipitation and temperature from the Water and Global Change (WATCH) data (Harding et al., 2011; Weedon et al., 2011). The authors reviewed

56 drought indicators and then tested the selected group to compare performance and analyse suitability at a global scale.

Among the variety of drought indicators, the Standardized Precipitation Index (SPI) is used most extensively because it only requires precipitation for its calculations. This information is generally available in any location via at-site observations or remote-sensing estimations. Another element that makes the SPI suitable is that results can be compared between different locations, which is important when conducting spatio-temporal analyses (Mckee et al., 1993). The following section introduces the SPI and other drought indicators. This review is not extensive but serves as a foundation for chapters 3, 4, 5 and 6, where drought indicators are either computed or used for drought monitoring and characterisation.

2.2.1 Meteorological drought indicators

Drought indicators aim to quantify meteorological drought. As mentioned, they can also be used to detect other types of drought. The main representative in this group of indicators is the SPI introduced by McKee et al. (1993). In this index, precipitation probability distribution – typically gamma distribution – is first computed, then equalled to the normal distribution (Figure 2.1). Thus, the SPI represents the value of the standard normal variable with a mean (μ) of zero and a standard deviation (σ) of unity.

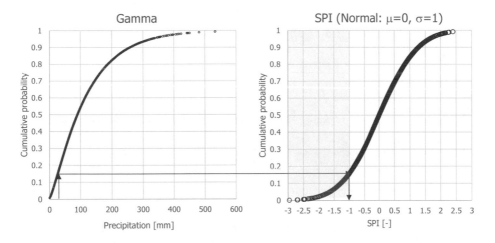

Figure 2.1 Original SPI calculation scheme proposed by McKee et al. (1993). First, precipitation is fitted by gamma distribution (left). Second, the cumulative probability of fitted precipitation is used to compute the SPI value considering normal distribution (right). Dry periods occur when the SPI is below 0. The threshold of SPI = -1 is often used to calculate drought (shaded area, right panel).

Different drought categories are distinguished according to the SPI's magnitude: moderate, severe or extreme (Table 2.2). Operationally, the intervals shown in Table 2.2

are useful to qualitatively indicate drought severity. For a more complete and quantitative description of drought, an SPI threshold is often used to calculate the drought's beginning and end (Figure 2.2). The onset occurs when the drought indicator is below the threshold, and the end occurs when it is above. The difference between the end and the onset is used to calculate duration. The sum of the magnitudes of drought indicators between the onset and end defines the severity or magnitude. Finally, intensity is computed as the ratio of drought magnitude and duration.

Table 2.2 Drought categories according to the SPI magnitude.

Drought indicator magnitude	Drought category
-1.5 < SPI ≤ -1.0	Moderate
-2.0 < SPI ≤ -1.5	Severe
SPI ≤ -2.0	Extreme

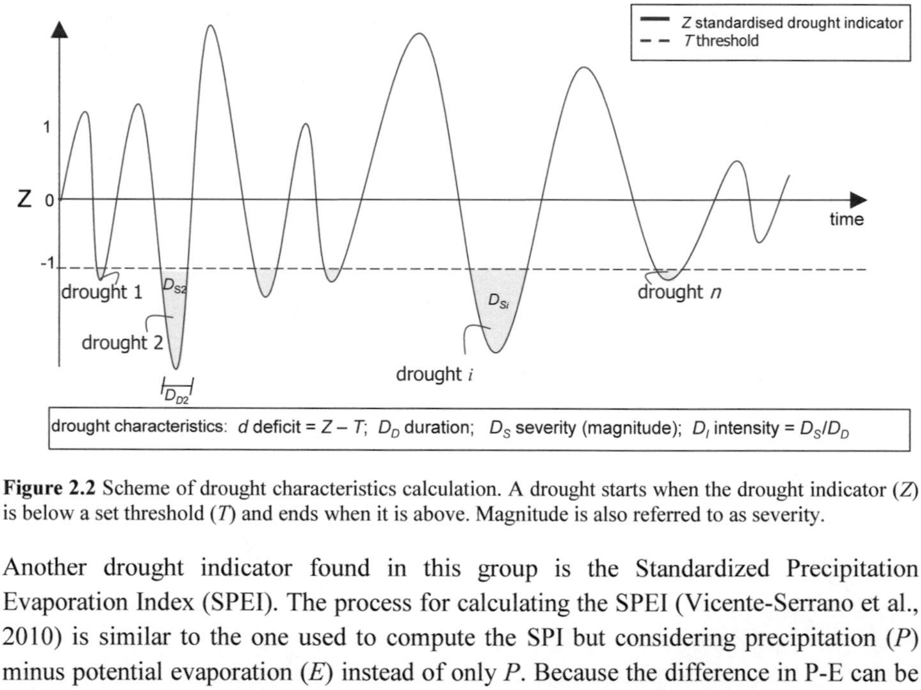

Figure 2.2 Scheme of drought characteristics calculation. A drought starts when the drought indicator (Z) is below a set threshold (T) and ends when it is above. Magnitude is also referred to as severity.

Another drought indicator found in this group is the Standardized Precipitation Evaporation Index (SPEI). The process for calculating the SPEI (Vicente-Serrano et al., 2010) is similar to the one used to compute the SPI but considering precipitation (P) minus potential evaporation (E) instead of only P. Because the difference in P-E can be negative, gamma distribution is not recommended for calculating the SPEI. Log-normal, generalised logistic or generalised extreme value distribution is preferable in this case. Several studies have tested the SPEI's suitability for agricultural and hydrological drought monitoring; its inclusion of evapotranspiration, for example, gives the SPEI a higher correlation with agricultural and hydrological drought indices (Bachmair et al.,

2015, 2016; Diaz-Mercado et al., 2016; Li et al., 2015; Naumann et al., 2014; Maskey and Trambauer, 2014; Vicente-Serrano et al., 2012).

2.2.2 Agricultural drought indicator

The procedure for calculating SPI can be extended to other variables. For agricultural drought, soil moisture is the variable to consider. This requires an empirical distribution approach because soil moisture does not fit well with gamma probability distribution. In such an approach, the data is first sorted into ascending order. The smallest value occupies the position i of 1 and the highest value the position of n (i.e. the number of observations). The empirical probability p is calculated using Eq. 2.1, which is the formation of Kaplan-Meier, though other expressions may be used instead.

$$p(x_i) = i/(n+1) \tag{Eq. 2.1}$$

Once the empirical probability p is calculated, the procedure follows that of the SPI in that the standardised value is obtained using the mean (μ) of zero and the standard deviation (σ) of unity (Figure 2.2, right). This procedure, including the empirical probability distribution, can also be used to calculate the SPI and SPEI.

Another agricultural drought indicator was proposed by Narasimhan and Srinivasan (2005). They introduced the Evapotranspiration Deficit Index (ETDI), which involves the water stress ratio defined by Eq. 2.2:

$$WS = (PET - AET) / PET \tag{Eq. 2.2}$$

where potential evapotranspiration (PET) and actual evaporation (AET) are the rates of monthly reference potential evaporation and monthly actual evaporation, respectively. WS values are used to calculate the monthly water stress anomaly (WSA) as:

$$WSA_{y,m} = (WS_{y,m} - MWS_m) / (MWS_m - minWS_m) \times 100, \text{ if } WS_{y,m} \leq MWS_m$$

$$WSA_{y,m} = (WS_{y,m} - MWS_m) / (maxWS_m - MWS_m) \times 100, \text{ if } WS_{y,m} > MWS_m \tag{Eq. 2.3}$$

where $MWS_{y,m}$ is the long-term median of water stress of month m, $maxMWS_m$ is the long-term maximum water stress of month m, $minWS_m$ is the long-term minimum water stress of month m and $WS_{y,m}$ is the monthly water stress ratio (y = 1968–2008 and m = 1–12). $MWS_m - minWS_m$ can be zero when using the long-term median, in that case, the long-term average can be used instead (Diaz-Mercado et al., 2016). Narasimhan and Srinivasan (2005) scaled the ETDI to between -4 and 4 to be comparable with the PDSI. Maskey and Trambauer (2014) proposed to scale the ETDI between -2 and 2 to make it comparable to the SPI and SPEI, as represented by:

$$ETDI_{y,m} = 0.5ETDI_{y,m-1} + (WSA_{y,m}/100) \tag{Eq. 2.4}$$

$$ETDI_{1,1} = WSA_{1,1}/100$$

2.2.3 Hydrological drought indicator

Standardized Runoff Index (SRI)

The SRI follows the same concept as the SPI (Shukla and Wood, 2008) but using runoff. In SRI calculation, the observed/simulated runoff time series is first fitted to a probability function. The cumulative probability is then translated into standardised normal distribution via the same procedure used to determine the SPI. Studies that applied this drought indicator include Barker et al. (2016), Diaz-Mercado et al. (2016) and Trambauer et al. (2014).

2.2.4 Main difficulties of a standardized drought indicator computation

Computing a standardised drought indicator requires handling several issues, such as fitting the most appropriate distribution function to the signal in question (i.e. P, D=P-PET or runoff). When using SPEI, the most suitable method estimation for PET is also an issue to consider.

Vicente-Serrano et al. (2010) proposed using three-parameter log-logistic distribution to fit D when formulating the SPEI. This method is based on data from eleven observatories around the world. However, some critics have pointed out that the chosen data may not completely represent global climate diversity (Beguería et al., 2014). To evaluate this possible limitation, a parameter-fitting review of the distribution function proposed by Vicente-Serrano et al. (2010) was performed using global gridded data from the Climatic Research Unit (CRU) TS3.10.01 dataset (Harris et al., 2014; http://badc.nerc.ac.uk) and three formulations for computing PET (i.e. Thornthwaite, Hargreaves and Penman-Monteith). Ultimately, the researchers approved the use of log-logistic distribution for fitting D. As for PET formulation, the SPEI was found to reveal only small differences in areas with high precipitation, while the index changed significantly in areas with low precipitation when different PET equations were selected.

Conversely, Stagge et al. (2015) recommend using generalised extreme value (GEV) distribution when conducting evaluations in Europe. They applied WATCH data (Weedon et al., 2011) to compute PET using the Penman-Monteith approach. Naumann et al. (2014) used log-logistic distribution to fit D in the context of Africa. Temperatures from ERA-Interim (Dee et al., 2011; section 2.2) were used to derive PET, per Thornthwaite's method. The authors computed D using different precipitation datasets to compare SPEI values over Africa. These datasets came from ERA-Interim and the Global Precipitation Climatology Project (GPCP, Huffman et al., 2009). The results suggest that most differences when computing drought indicators arise from uncertainties in precipitation datasets rather than distribution parameter estimations.

Standardised drought indicators based on the SPI methodology are attractive for drought analysis because their values can be compared or correlated, not only among different indicators but also among different locations around the globe. However, there is a need for a generalised framework for computation purposes, particularly when calculating distribution function to assess droughts based on different hydro-meteorological variables.

2.3 DROUGHT CALCULATION

One of the first studies to consider the spatial extent (area) of drought was the work of Yevjevich (1967). He viewed drought as a phenomenon with a territorial extension and calculated time series of drought areas using precipitation data. These time series were also used to define the onset, end and duration of each drought. In the absence of observed precipitation, the methodology was applied using synthetic data. Later, Yevjevich and Karplus (1973) expanded the original methodology to apply precipitation data in two regions of the United States. Here, spatial extent was calculated using Thiessen polygons. The authors proposed the use of a grid system to represent and handle data, but it was not implemented. In such a grid system, information would be arranged in a matrix of columns and rows, with each cell representing a geographical location. At each time step, the information represented in each cell can change.

Following the work of Yevjevich and Karplus (1973), Tase (1976) calculated drought areas using information arranged in a grid system (i.e. grid data). Due to limited availability of stations records, a technique based on the Monte Carlo simulation was applied to generate synthetic information. Later, Bhalme and Mooley (1980) developed an application for computing drought areas using grid data interpolated from stations. They used precipitation data to conduct drought analyses monthly in India. The observed droughts were categorised from the most severe (the largest average area) to determine the worst drought years. In Europe, Zaidman et al. (2002) estimated drought areas using rainfall and runoff information, spatially interpolated their data from stations. They used time series of drought areas to calculate the phenomenon's onset, end and duration.

These studies all viewed the entire regional drought area under analysis as the spatial extent of the phenomenon. However, Andreadis et al. (2005) proposed a methodology where the area is not considered as a whole, but rather a contiguous portion defined in space and time. Through a clustering technique applied to the grid data, they calculated what are called 'drought events'. Each calculated drought event has a duration and an area. Similarly, Corzo Perez et al. (2011) proposed a methodology in which the centroid of the cluster defines the geographic location of the drought. This methodology may be applied using global simulated runoff data. Finally, Lloyd-Hughes (2012) applied a technique based on that by Andreadis et al. (2005) to calculate drought events in Europe using precipitation data. Each drought event was defined in three dimensions: duration, location and volume (number of cells).

Herrera-Estrada et al. (2017) presented their own analysis of how droughts move in space. They perform spatial tracking by calculating the displacement between consecutive areas in time, which allows tracks to be identified. Diaz et al. (2018) presented an alternate methodology to build the spatial path of a drought. Per their view, a path is defined as successive spatial tracks of a drought. Such a path should be calculated to define the onset and end of each drought in time and space. The authors suggest that the information provided by these drought path calculations can be used to build a model that will help to predict droughts, particularly in terms of location and spatial extent.

All these prior studies demonstrate how drought characterisation (i.e. the calculation of such features as severity, duration, spatial extent and location) has been improving as new data and advances in spatio-temporal analysis become available.

2.4 VISUALISATION AND ANALYSIS OF DROUGHT VARIATION

Data visualisations are widely used for drought analysis. Selected papers perform this analysis explicitly in terms of spatial extent (area), which is one of the target variables accounted for in the present study. The example studies described below identify drought according to different parts of the water cycle (e.g. precipitation, soil moisture, and runoff) and by following various calculation approaches.

Visual approaches have been conducted using various applications. Zaidman et al. (2001) analysed drought development in Europe by applying a three-step procedure involving precipitation and streamflow. First, water anomalies were calculated over time series of grid data. The grid data were obtained by interpolating records from stations throughout Europe. Water anomalies were calculated using a standardised drought index. Second, a threshold was applied to indicate drought and non-drought conditions in each cell of the grid for each time step. Finally, the spatial extent (area) of drought was calculated in each time step, with all cells in the drought deemed part of the same drought area. Zaidman et al. (2001) were able to calculate the drought areas for specific well-known historical droughts in Europe and identify the worst ones (i.e. those with greatest extents) on maps. They used line charts to show how drought areas change over time; these charts helped interpret drought development in Europe and its subregions between 1975 and 1990.

Similarly, Peters et al. (2006) analysed how drought propagates from precipitation to the groundwater system. To do so, they computed drought duration using spatially distributed simulations of recharge and hydraulic head. They presented maps of the spatial distribution of drought duration for selected years. Using area charts, they showed drought areas at specific times during the analysis period. The maps and charts aided the researchers in concluding that a drought caused by precipitation is attenuated and delayed in the groundwater system. To calculate drought, Peters et al. (2006) used the threshold level approach.

Other studies that applied the use of line and area charts were conducted by Andreadis et al. (2005), Corzo Perez et al. (2011), Diaz et al. (2019), Hannaford et al. (2011), Lloyd-Hughes (2012), Justin Sheffield et al. (2009), Tallaksen and Stahl (2014), Van der Schrier et al. (2013) and Van Huijgevoort et al. (2013).

The third type of visualisation, similar to the line (area) chart, is the bar chart. Examples include the works of Bhalme and Mooley (1980), WMO (2006) and Zaidman et al. (2001). Bar charts can be horizontal or vertical, though the latter is often preferred.

Data in line, area and bar charts is usually presented along the x-axis to display the information over time. This layout allows the higher-than-average data to be identified within the analysis period. The layout also allows any trends in the reduction or increase of drought characteristics to be easily detected visually. However, these visualisations are limited in that drought patterns, such as seasonality, are difficult to identify. Seasonality refers to the situation in which drought characteristics show regular and repetitive changes in almost similar periods over time in a calendar year.

Heat maps, in contrast, are data visualisations that aid recognising drought seasonality. Applications of heat maps, or colour-coded tables, include the works of Prudhomme and Sauquet (2007), who analysed drought changes over France; Hannaford et al. (2011), who presented a drought analysis of Europe; and Corzo Perez et al. (2011) and Van Huijgevoort et al. (2013), who performed global assessments of spatio-temporal drought development.

Heat maps are also used for the visual representation and analysis of correlations among different droughts characteristics. Characteristics are usually calculated using drought indicators, hydro-meteorological variables or aggregation periods. Heat maps are often used in drought monitoring where two or more drought characteristics are compared to find correlations between them; for example, runoff-derived drought intensity might be compared with that derived from precipitation. Such a comparison would offer information about any good proxy of the runoff-derived drought using only meteorological variables, such as precipitation, which are easy to gather. In heat maps, values of correlation coefficients between drought characteristics are represented with colours. The highest values are usually indicated with a specific colour. These coloured spots (cells) help to reveal the most promising proxies of runoff-derived drought.

Another type of visualisation used for drought analysis is the radial chart. Its application is found in studies where the development of more than two drought characteristics are analysed. For instance, Lloyd-Hughes et al. (2010) studied the occurrence and spatial extent of drought in Europe. They computed the area, onset and duration of each calculated droughts. In the constructed radial histograms, which consisted of coloured radii, area magnitude was presented in colour-coded intervals; onset was indicated by the position of the radius in the chart; and duration was represented by the length of the radius. This layout allowed the researchers to identify drought-rich periods over the year, as well as average duration and the month in which droughts tend to start.

Another example is the work of Lloyd-Hughes (2012), who used precipitation data to calculate drought areas across Europe. In this study, the centroid of the drought area represented its geographical location. These locations helped Lloyd-Hughes to determine the main routes followed by droughts. Three drought characteristics were plotted on radial charts: duration, area and main location (north, south, west and east). Much like Lloyd-Hughes et al. (2010), Lloyd-Hughes used colours to encode area magnitude. The length of the radius represented duration, and the position of the radius in the diagram indicated drought area location.

The use of radial charts to analyse drought is limited to a few applications (de Brito, 2021; United Nations Office for Disaster Risk Reduction, 2021). In hydrology and related areas, including drought research, radial graphs are mainly used to compare the performance metrics of models or algorithms. These graphs are known as spider charts (e.g. Smith et al., 2019). The use of radial charts is much more widespread in other science branches. In genetics, radial charts often appear in analyses of complex relationships concerning genes; specialised software exists for generating such visualisations. Ciros (Krzywinski et al., 2009), for example, is a computer package that allows relationships between different entities to be explored. Based on these applications, it is expected that drought research would benefit from incorporating radial charts into analyses of spatio-temporal drought patterns.

2.5 MACHINE LEARNING TECHNIQUES IN DROUGHT STUDIES

Machine learning (data-driven) models involve assessing mathematical equations to model a phenomenon using only data, as opposed to process (physically based) models describing the essence of physical processes, e.g. water motion or heat fluxes. The use of ML models has become increasingly common in recent decades (Chlingaryan et al., 2018; Rahmati et al., 2020; Solomatine and Ostfeld, 2008; Udmale et al., 2020; van Klompenburg et al., 2020).

Typically, a data-driven model can be defined based on the connections between system state variables (input, internal and output variables), with only a limited number of assumptions about the physical behaviour of the system. Contemporary data-driven models can go much further than those used in conventional empirical hydrological modelling. Newer models allow for solving numerical prediction problems, reconstructing highly nonlinear functions, performing pattern recognition and classification, and building rule-based systems (Solomatine and Ostfeld, 2008). Such models assume the presence of a considerable amount of data to describe the modelled system's physics, i.e. meteorological or hydrological phenomena, as well as spatial data coming from remote sensing.

In drought calculation and monitoring, Mallya and Tripathi (2013) note that researchers have addressed the topic using different ML models, such as nonlinear regression models, hybrid models and artificial neural networks (e.g. Shin and Salas, 2000; Kim and Valdes, 2003; Mishra et al., 2007). Advanced statistical techniques have also been utilised. Copulas have been applied to model the joint dependence structure of drought characteristics (e.g. Wong et al., 2010, 2013; Madadgar and Moradkhani, 2013). Kao and Govindaraju (2010) suggest using a joint deficit index, which applies empirical copulas to provide a probability-based description of the overall drought status.

Hao and Singh (2012) propose using entropy theory to construct the bivariate joint distribution of drought duration and severity and to make comparisons with a copula-based analysis. Farahmand and AghaKouchak (2015) , meanwhile, recommend a generalised framework for deriving nonparametric univariate and multivariate standardised indices, e.g. SPI. They argue that current indicators suffer from deficiencies, including temporal inconsistency and statistical incomparability, because such indicators rely on a representative parametric probability distribution function that fits the data. The framework they propose draws on different variables, such as precipitation, soil moisture and relative humidity, whilst using an empirical plotting position without having to assume representative parametric distributions.

In the calculation of drought as an event with a spatial extent, ML models have been used mainly for calculating contiguous drought areas (Andreadis et al., 2005; Corzo Perez et al., 2011; Herrera-Estrada et al., 2017; Lloyd-Hughes, 2012; Sheffield et al., 2009; Van Huijgevoort et al., 2013; Vernieuwe et al., 2019). This topic is addressed in detail in Chapters 5 and 9.

Different clustering techniques have been used to identify drought events. These events are defined by a spatial extent, onset and end in space and time, duration and intensity. Some applications have even proposed the centroid of drought clusters as the location of the events (e.g. Herrera-Estrada et al., 2017). Drought trajectory calculation, analysis and characterisation are topics that have been addressed in recent years.

There are several examples of ML models in drought prediction. In the 'Special issue on data-driven approaches on droughts' edited by Govindaraju (2013) , there are cases on this subject. In general, ML models were tested mainly to predict different types of drought indices. Regarding predicting drought impacts, for example, in agriculture, ML models have been widely tested. The works of Chlingaryan et al. (2018) and van Klompenburg et al. (2020) show reviews of different types of ML models and inputs used to predict crop yield, a variable commonly used in agricultural assessments. Both reviews show that amongst the most used inputs are drought index values. However, variables related to the spatial characterisation of drought, such as drought area, are not reported in these reviews.

2.6 SUMMARY AND CONCLUSIONS

After reviewing previous studies, this research focuses on the use of standardised drought indicators for the calculation of drought for two reasons. First, to assess drought characteristics, a generic methodology for drought identification is required to resemble hydrological cycle components (i.e. precipitation, precipitation-evaporation and runoff) (Chapter 6). Second, the spatial component of drought analysis makes it convenient to work with standardised drought indicators in order to compare different parts of the same region under study (Chapters 5 and 9).

The literature review has identified knowledge gaps in previous research. These gaps formed the basis for formulating the objectives of this study presented in Chapter 1.4.

Recent research points out that characterising drought in space and time requires further exploration. The use of spatio-temporal approaches, which provide a more accurate onset, duration and intensity, is expected to result in better characterisations of drought events. The literature review indicates that ML techniques can help calculate drought events considering the spatial context, i.e. calculation of drought areas and clusters.

The literature review also shows a gap in methodologies to characterise the spatio-temporal development of drought. Methods that help in the calculation of spatial drought trajectories are lacking.

Visual approaches are likewise needed to analyse the variation in spatio-temporal drought characteristics. Radial and polar charts that help manage long data periods and detect spatial patterns can be explored in the analysis of droughts.

Finally, the use of the spatial properties of drought, such as drought area, has not been fully explored in predicting drought impacts.

3

METHODOLOGICAL FRAMEWORK

The following paragraphs provide an overview of the general methodological framework used in this research, with references to other chapters where particular methods are presented in detail. This methodology was designed to achieve each of the objectives of this research formulated in Chapter 1. Figure 3.1 shows how the objectives and the different chapters of this paper are linked. The figure also indicates where to find the results of the methodology's implementation for each objective.

The literature review presented in Chapter 2 gives an overview of the current methods and concepts related to drought characterisation. The literature review allowed for pointing at the appropriate methods to be used in the methodology towards achieving the research objectives. Case studies were chosen with the objectives in mind, and are presented in Chapter 4.

Chapter 5 describes the proposed approach to characterising drought dynamics in space and time. This approach was developed as part of the first objective (O1). This approach was then applied to all the objectives, according to the particular needs of each.

The final chapter presents conclusions and recommendations based on the results of the methodology's application.

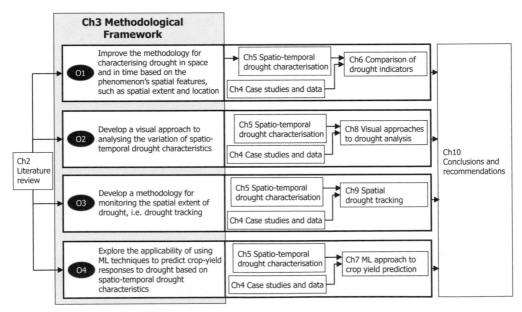

Figure 3.1 Schematic overview of PhD research methodology showing the link between objectives (O1 to O4) formulated in Chapter 1, and the dissertation chapters (Ch2 to Ch10).

3.1 IMPROVE THE METHODOLOGY FOR CHARACTERISING DROUGHT IN SPACE AND TIME BASED ON THE PHENOMENON'S SPATIAL FEATURES, SUCH AS SPATIAL EXTENT AND LOCATION

Figure 3.2 shows the methodology for meeting the first objective (O1). A review of the methods used to calculate and characterise drought was first carried out to develop this section's methodology. Based on current developments, an approach was built to characterise drought dynamics in space and time. This approach is described in detail in Chapter 5.

As the first application, different drought indicators were compared and detailed in Chapter 6. First, drought indicators were selected to calculate drought. Types of indicators and their descriptions are shown in Chapter 2. An additional indicator based on evapotranspiration was developed for the reasons stated in Chapter 6. Drought indicators were then calculated for a case study. Afterwards, the methodology described in Chapter 5 was applied to calculate the spatio-temporal characteristics of drought, such as spatial extent, magnitude and duration. The results are presented in Chapter 6. Figure 1 shows the steps to carry out the development of the drought indicator for hydrological drought analysis at a large scale. An ML model was used to either compute the drought indicator or carry out the drought characterisation.

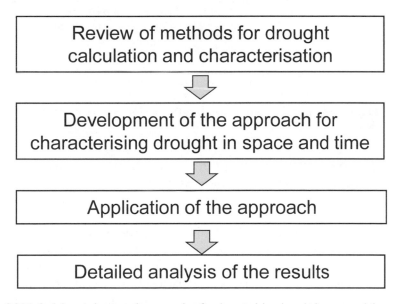

Figure 3.2 Methodology to improve the approaches for characterising drought in space and time.

25

3.2 DEVELOP A VISUAL APPROACH TO ANALYSING VARIATIONS OF SPATIO-TEMPORAL DROUGHT CHARACTERISTICS

Visual approaches for analysing drought variations were developed to study the characteristics of drought calculated using the method presented in Chapter 5 (Figure 3.3). These characteristics included spatial extent, the number of clusters and the mean distance between clusters. Changes of these characteristics over time allowed for identifying different patterns, such as seasonality and persistence, to be described (see Chapter 5). Based on the different visual approaches used for drought analysis (see Chapter 2), radial and polar graphs were designed. In the designed charts, the different characteristics of droughts were encoded (i.e. displayed) through colours and other characteristics to visually identify drought variations and patterns. The results of applying these graphical approaches are shown in Chapter 8.

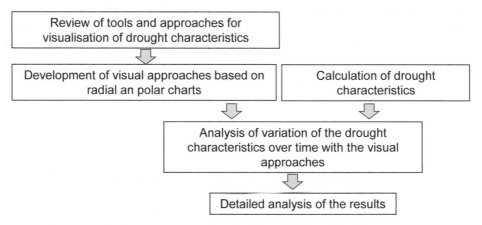

Figure 3.3 Methodology to develop visual approaches to analysing variations of spatio-temporal drought characteristics.

3.3 DEVELOP A METHODOLOGY FOR MONITORING THE SPATIAL EXTENT OF DROUGHT, I.E. DROUGHT TRACKING

The methodology for developing an approach to monitor and track a drought's spatial extent was as follows (Figure 3.4): First, a literature review of the methods for calculating spatial extent and other related characteristics, such as location, was carried out. Chapter 5 illustrates this step. Subsequently, a method was developed to calculate the spatial trajectory of a drought area. The method was then applied in a case study for which the trajectories of large drought events were identified. The application of the methodology for monitoring droughts is described in Chapter 9. Chapter 10 presents conclusions and recommendations. An approach to building a model for predicting the spatial extent and location of a drought is also introduced in Chapter 10. This new approach to tracking drought areas is expected to aid in drought monitoring and prediction.

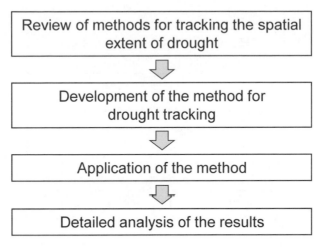

Figure 3.4 Methodology for developing an approach to monitoring the spatial extent of drought (i.e. drought tracking).

3.4 EXPLORE THE APPLICABILITY OF USING ML TECHNIQUES TO PREDICT CROP-YIELD RESPONSES TO DROUGHT BASED ON SPATIO-TEMPORAL DROUGHT CHARACTERISTICS

Droughts have diverse impacts on human activities and the environment. For example, in agriculture, drought causes quantitative economic losses due to crop-yield losses (see Chapter 7 for details). Decline in crop yield can also compromise food supplies, putting many human lives at risk, as well as incurring significant economic losses when crop distribution is affected.

This research analyses the impacts of drought on agriculture because this area is one of the most affected by the phenomenon. The analysis focuses on crop yield because this measure is one of the most used and reported to quantify agricultural production. Chapter 7 presents the application and results of the methodology shown below. The reason for using artificial intelligence (ML) models instead of those based on crop growth is explained in Chapter 7.

The crop-yield prediction model was constructed as follows (Figure 3.5): First, the main ML models used for drought prediction were reviewed. This literature review is shown in Chapter 2. An ML approach was then designed to predict crop yield using two of the most used ML models (see Chapter 2). Drought spatial extent was considered as input data for the ML models. This characteristic was calculated according to the methodology presented in Chapter 5. The results are presented in Chapter 9.

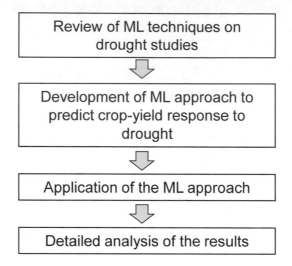

Figure 3.5 Methodology to predict crop-yield responses to drought based on spatio-temporal drought characteristics and ML techniques.

4

CASE STUDIES AND DATA

This chapter is partially based on the following publications.

Diaz, V., Corzo Pcrez, G. A., Van Lanen, H. A. J., Solomatine, D., and Varouchakis, E. A. (2020). An approach to characterise spatio-temporal drought dynamics. Advances in Water Resources, 137, 103512. https://doi.org/10.1016/j.advwatres.2020.103512

Diaz, V., Corzo Perez, G. A., Van Lanen, H. A. J., Solomatine, D., and Varouchakis, E. A. (2019). Characterisation of the dynamics of past droughts. Science of The Total Environment, 134588. https://doi.org/10.1016/j.scitotenv.2019.134588

Diaz, V., Corzo, G., Lanen, H. A. J. Van, and Solomatine, D. P. (2019). 4 - Spatiotemporal drought analysis at country scale through the application of the STAND toolbox. In G. Corzo and E. A. Varouchakis (Eds.), Spatiotemporal Analysis of Extreme Hydrological Events (pp. 77–93). Elsevier. https://doi.org/10.1016/B978-0-12-811689-0.00004-5

4.1 MEXICO

4.1.1 Large scale

Mexico is a country prone to drought that has serious adverse impacts on human lives, and was chosen as a case study. This work analyses drought variation at a country scale using visual approaches based on radial charts presented in Chapter 8. The data used in the present study were retrieved from the Self-calibrating Palmer Drought Severity Index (scPDSI) for Global Land (scPDSIcru global, https://crudata.uea.ac.uk/cru/data/drought/#global). The scPDSIcru is based on the PDSI introduced by Palmer (1965) (Chapter 2), but incorporates the methodology presented by Wells et al. (2004) to calibrate the water-budget system (i.e. scPDSI), and a snowmelt model to extent its use to colder regions (Van Der Schrier et al., 2013). These additions allows scPDSI be used for drought analysis at large scale. An example of the use of the scPDSI in a global analysis is the work of Van Der Schrier et al. (2013). The spatial resolution of scPDSIcru is 0.5 degrees and it was calculated on a monthly basis.

4.1.2 Catchment scale

Hydro-meteorological data from the La Sierra River basin (Mexico) were used to compare different drought types and indicators (Figure 4.1). For this basin, Diaz Mercado et al., 2015) calculated runoff and evapotranspiration via hydrological modelling. They reported that model inputs included daily historical weather data from 40 weather stations (1968 to 2008) of both precipitation and temperature. The basin was divided into squares of 5 x 5 km. Then, precipitation and temperature values were interpolated spatially by Inverse Distance Weighting (IDW). Finally, the hydrological model was calibrated and validated using daily streamflow at a gauging station located at the outlet. Streamflow data was collected from 1968 to 1999. The overall Nash-Sutcliffe error (NSE) was 0.87, and the coefficient of determination (R^2) value was 0.87. The outputs of the hydrological simulations, such as evaporation and runoff, were used in this research to analyse the performance of different droughts indicators (Chapter 6).

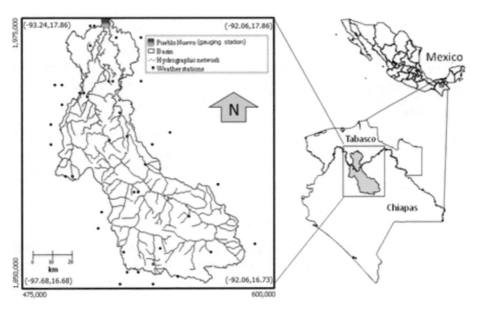

Figure 4.1 La Sierra River basin located at the southeast of Mexico. The outlet is in the north of the basin (Diaz Mercado et al., 2015).

4.2 INDIA

4.2.1 Large scale

This research characterises drought dynamics in the 'real-world' case of India, which regularly suffers from drought (Chapter 9). In principle, any other region would also qualify. The SPEI was selected for this study, but any other drought indicator would also suffice. Data from the SPEI Global Drought Monitor (http://spei.csic.es/) was used for drought tracking and to characterise drought dynamics. The chosen methodology for calculating the SPEI was similar to that used with the SPI proposed by Mckee et al. (1993), but with added consideration for the difference between precipitation (P) and potential evaporation (E). SPEI data from the drought monitor were charted in grid form for different temporal aggregation periods. SPEI-6, which corresponds to six-month P − E accumulation anomalies, was used. This aggregation usually refers to extended periods of lack of water availability; this indicates that the consequences of meteorological drought are similar to those caused by hydrological drought (World Meteorological Organization (WMO), 2012). In this study, SPEI-6 was selected as the drought indicator to characterise past droughts. Analyses were conducted monthly. The spatial resolution of the SPEI-6 data is 0.5 degrees.

31

4.2.2 Regional scale

A ML approach was developed to predict crop yield using percentages of drought area as input (Chapter 7). This ML approach was applied in three regions of India: Bihar, Odisha and West Bengal (Figure 4.2).

Crop yield

Rice is the most important food grain in East India, so it was selected to assess ML-oriented crop-yield predictions. Rice from this region accounts for roughly 85 percent of the total rice production in India (Ghosh et al., 2014). ML models were constructed for the states of Bihar, Odisha and West Bengal (Figure 4.2). State-wise crop-yield data were retrieved from 1966 to 2015 (49 years) through the Indian Directorate of Economic and Statistics from the Department of Agriculture (DAC) (http://eands.dacnet.nic.in/).

There are three crop seasons in India: Rabi, Kharif and Zaid. Of these, the Kharif season was chosen for study because it is the largest in terms of crop production. Kharif crops are sown in June and harvested in November/December. Seasonal crop-yield data was obtained from the DAC website and arranged into time series per region. One value was assigned to each year of crops harvested in the Kharif season.

Two important clarifications have to be made. First, in late 2000, Bihar was divided into two states: Bihar and Jharkhand. Thereafter, rice data was reported separately. In this study, both states are marked as the 'Bihar region'; the crop-yield data from 2000 to 2015 is the reported sum of current Bihar and Jharkhand. Second, in 2011, Orissa was renamed Odisha, but the territory remains the same. In this case, crop yield data for Odisha is that reported for the former Orissa and the current Odisha.

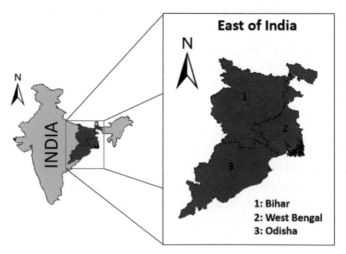

Figure 4.2 Case study location.

Drought indicator

Soil moisture is the preferred variable for calculating agricultural drought indicators. However, another widely disseminated way to indirectly infer a drought indicator is to use meteorological drought indicators as proxies. Among these, the SPEI has shown to be useful in assessing agricultural drought. The SPEI follows a similar methodology as that of the widely used SPI, but with added consideration for the difference between precipitation and evapotranspiration. SPEI data was retrieved from the SPEI Global Drought Monitor (https://spei.csic.es) between 1901 and 2015. The spatial resolution of the drought indicator data is 0.5 degrees. The SPEI data was available at different aggregation periods; for this study, it was retrieved for the aggregation periods of 1, 3, 6, 9 and 12 months, indicated as SPEI1, SPEI3, SPEI6, SPEI9 and SPEI12, respectively (Chapter 7).

5

SPATIO-TEMPORAL DROUGHT CHARACTERISATION

This chapter is partially based on the following publications.

Diaz, V., Corzo Perez, G. A., Van Lanen, H. A. J., Solomatine, D., and Varouchakis, E. A. (2020). An approach to characterise spatio-temporal drought dynamics. Advances in Water Resources, 137, 103512. https://doi.org/10.1016/j.advwatres.2020.103512

Diaz, V., Corzo, G., Lanen, H. A. J. Van, and Solomatine, D. P. (2019). 4 - Spatiotemporal drought analysis at country scale through the application of the STAND toolbox. In G. Corzo and E. A. Varouchakis (Eds.), Spatiotemporal Analysis of Extreme Hydrological Events (pp. 77–93). Elsevier. https://doi.org/10.1016/B978-0-12-811689-0.00004-5

Corzo, PGA, Diaz, V., Laverde, M. (2018). Spatiotemporal hydrological analysis. International Journal of Hydrology, 2(1):25-26. doi: 10.15406/ijh.2018.02.00045

5.1 INTRODUCTION

It has been highlighted that improved drought analyses would enable the development and implementation of more successful national policies to mitigate the negative impacts of drought (WMO, 2006). The WMO also points out that to reduce these negative impacts, new technologies and methods to improve drought characterisation must be developed.

The objective of this chapter is to introduce the Spatio-Temporal ANalysis of Drought (STAND) and illustrate its use through case studies. The use of STAND method is illustrated in the following chapters through different case studies.

5.2 SPATIO-TEMPORAL ANALYSIS OF DROUGHT (STAND)

The drought analysis proposed here can be performed at different levels, from the most general (e.g. drought areas) to more detailed (e.g. contiguous areas in space or clusters). The aim is always to describe how drought evolves in space and changes over time.

5.2.1 Temporal analysis

Figure 5.1 shows a time series of a DI whose values oscillate from -3 to 3. The negative values are associated with drought anomalies. Each cell in the grid data has a DI time series. The characterisation of time series events is carried out for each cell. Per McKee et al. (1993), a time series event starts at the time ts, when the DI value is below a set threshold (T), and ends at the time te, when the DI is above it. The duration (d) and deficit (df) of each i-th time series are computed using Eq. 5.1 and 5.2, respectively.

$$d_i = te - ts \qquad\qquad\qquad\qquad\qquad\qquad\text{(Eq. 5.1)}$$

$$df_i = \sum_{t=ts}^{te}(DI(t) - T) \qquad\qquad\qquad\qquad\text{(Eq. 5.2)}$$

The deficit is standardised and expressed as a percentage using Eq. 5.3:

$$dfs_i = 100 \times df_i / \bar{x} \qquad\qquad\qquad\qquad\text{(Eq. 5.3)}$$

where dfs_i is the standardised deficit of the i-th time series, and $\bar{x}$ is the mean of the deficit values of the analysed time series.

To summarise the d and df computations over each time series, their median values are calculated in each cell. In this way, maps of the spatial distribution of d and df are obtained.

5.2.2 Spatio-temporal analysis: first approach

This procedure follows that defined by Corzo Perez et al. (2011). To account for how much the spatial coverage of drought is changing, the amount of area affected in each time step must be calculated. The DI values must be converted into events (binary representation) based on a feasible threshold. This, threshold method is well-known, and its algorithm is described in Eq. 5.4. At each time step (t), 1s and 0s are used to indicate whether a cell is in drought or not (Ds, drought state). In each cell, if the DI value is below a set threshold (T), it is assigned the value of 1; otherwise, it is assigned as 0 (Eq. 5.4).

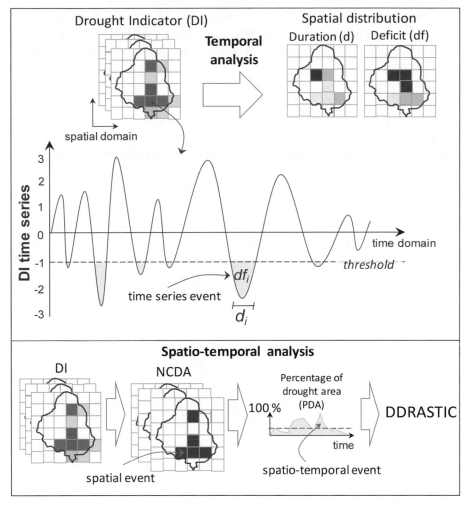

Figure 5.1 Schematic overview of the methodologies for drought analysis in STAND toolbox: Non-Contiguous Drought Area (NCDA) analysis, and Drought DuRAtion, SeveriTy, and Intensity Computing (DDRASTIC).

$$D_s(t) = \begin{cases} 1 \; if \;\; DI(t) \leq T \\ 0 \, if \;\; DI(t) > T \end{cases} \tag{Eq. 5.4}$$

DI is then converted into a binary representation of the spatial coverage of droughts. For a region, the percentage of drought area (PDA) is calculated using the area of the cells in drought and the region's area (Eq. 5.5). This allows the spatial variations in a time series spectrum to be represented. This process is implemented in the toolbox. Extensions to the method will be implemented later to create a non-binary representation that may better integrate droughts and provide more insight into the relative differences of the index values among the cells:

$$PDA(t) = 100/A_{tot} \cdot \sum_{c=1}^{N}(D_s(t) \cdot A) \tag{Eq. 5.5}$$

where A is the area of the cell c, and A_{tot} is the region area. Per this methodology, drought magnitude is the PDA value. The PDA time series allows the large droughts across a region to be identified. Small areas are excluded by applying a second threshold.

Drought DuRAtion, SeveriTy and Intensity Computing (DDRASTIC)

DDRASTIC was developed in this research and is based on the one proposed by McKee et al. (1993), but is extended for use with the PDA concept proposed in NCDA (Corzo Perez et al., 2011). Computing drought duration (DD) starts at the time step when the PDA is below the defined threshold. In a PDA series, this threshold is estimated using the 90th percentile of low PDA. The spatio-temporal event starts when the PDA is below the set threshold (T_{PDA}) and ends when it is above it (Eq. 5.6):

$$DD_j = t_E - t_S \tag{Eq. 5.6}$$

where j is the j-th spatio-temporal event, and DD, t_S and t_E are the drought duration, start and end time of the event j.

The region under the PDA curve represents drought severity (S, expressed as a percentage). This value is a measure of drought magnitude. S is calculated for each event j using Eq. 5.7:

$$S_j = \sum_{t=t_S}^{t_E} PDA(t) \tag{Eq. 5.7}$$

The intensity (I) is calculated as the ratio between S and DD (Eq. 5.8):

$$I_j = S_j/DD_j \tag{Eq. 5.8}$$

This I_j ratio can be interpreted as the mean value of PDA during the time DD_j. DD, S and I are calculated over the entire time series of PDA. Per this methodology, each triplet DD, S and I are characteristic of the so-called spatio-temporal event. It is important to highlight that if the spatial coverage is not large enough at any given time, there is no drought.

5.2.3 Spatio-temporal analysis: second approach

In this second approach, whether or not drought areas are connected in space matters. Drought units (clusters) are identified using the Contiguous Drought Area (CDA) analysis method. The CDA is composed of neighbouring cells in a drought. The conditions of drought or non-drought are indicated by 1s and 0s, respectively, in each cell. Simply put, drought conditions are indicated when the drought indicator is below or equal to a threshold. DIs are mathematical representations of a water anomaly (see 2.3 Section). In general, a CDA can be applied over any DI in grid form. These CDAs are computed per the CDA methodology for each time step.

The CDA analysis follow a connected-component labelling approach to cluster the cells in a drought (Haralick and Shapiro, 1992). Per this approach, a two-scan algorithm is applied. First, each cell is numbered for location issues. Then, an initial run is performed in which the binary grid is explored, and connected (contiguous) components (cells) are assigned provisional labels. These labels identify every cell's connection with its eight nearest neighbours, as in a grid section of 3 x 3 cells (nine cells in total), the central cell had eight surrounding cells. In this first run, the cell's label does not yet refer to the number of clusters, but to the cells with which the given cell is connected. A second scan is then carried out to identify similar cell connections (i.e. clusters). In this second run, clusters are indicated with a unique label. The grid could is examined indistinctly by columns or by rows. CDA analysis is conducted for each time step over the whole grid. For more details on CDA analysis, please refer to Corzo Perez et al. (2011).

The use of CDA relies on the assumption that the binary description of drought condition (0s and 1s) is homogeneous over the whole grid. If two or more cells denote drought (value of 1) conditions and are contiguous in space, they can be assumed to be part of the same drought unit. In this respect, it is recommended to choose a drought indicator that considers the normalisation of the spatial domain values.).

After computing the clusters (areas in drought), the most major (largest) one can be identified in each time step t and linked to calculate the drought trajectory. The tracking algorithm developed in this research is presented in Chapter 9, it focuses on calculating the major spatial drought extent in each time step and building the spatial trajectories. Small or one-cell units are excluded with the selection of the largest one, allowing the elimination of possible artefact drought areas.

Centroid localisation

After the major drought cluster is identified, its centroid (p) is calculated for each time step. The centroid is used as cluster's location because it resembles results presented by Corzo Perez et al. (2011) and Lloyd-Hughes (2012). Lowest drought indicator values can also be used to indicate the location of a given cluster (Andreadis et al., 2005; Herrera-Estrada et al., 2017). However, the centroid is chosen for this research because the drought indicator's spatial representation had already been reduced to 1s and 0s (i.e. drought and non-drought conditions, respectively).

Distance between drought clusters

The method presented for cluster calculation is restricted to the threshold selected for identifying drought conditions (i.e., cells with 0s and 1s). In this study, and the average distance between clusters, denoted by *lc*, was introduced to account for cases in which drought clusters are spatially close but not connected. The reasoning for using *lc* is as follows: if the average distance is small, clusters are more likely to be part of the same, large, quasi-contiguous one. Conversely, if the distance is large, it is more likely that the clusters are not spatially related.

The average distance between clusters (*lc*) is calculated using Eq. 5.9 for each time step.

$$lc(t) = \sqrt{\frac{1}{nc-1}\sum_{i=1}^{nc-1} l\min_i^2}$$
(Eq. 5.9)

To apply Eq. 5.9, the cluster with the largest area is first identified. Next, the minimum distance from the largest cluster is calculated for each of the remaining clusters. The number of clusters is denoted by *nc* in Eq. 5.9. The minimum distance (*l*min) is calculated for each of the *nc*-1 pairs of clusters. Each pair consisted of an *i*-th cluster and the largest one. Finally, the root mean squared distance (RMSD) of all distances is computed (*lc*) with Eq. 5.9. RMSD was chosen because of its sensitivity to higher and smaller distances, which in the identification of quasi-contiguous clusters is important. When high distances abound, RMSD results in a larger value than the mean distance; opposite happens when there are more small ones in which RMSD results in a lower value.

After completing the CDA analysis, the output was found to be the time series of the number of clusters (*nc*), the area of each cluster (DA) and the average distance (RMSD) between the clusters (*lc*).

5.2.4 Spatio-temporal drought patterns

'Drought pattern' refers to the way a drought exhibits occurrence and development in a spatial and temporal context. Several approaches can be applied to analyse drought patterns. Table 5.1 lists and describes the most common patterns. Drought pattern analyses can be conducted using the visual approaches described in Chapter 8. This list of patterns is not exhaustive, but rather serves as a basis towards a more complete one.

Table 5.1 Spatio-temporal drought patterns. DA, *nc*, and *lc* stand for drought area, number of clusters, and average distance between clusters, respectively.

Drought pattern	Description	Drought characteristic(s) that help(s) in its identification
Periodicity	The drought characteristic shows regular intervals of time in its occurrence.	DA
Seasonality	At fixed intervals of time in a calendar year, the drought characteristic shows low or high values. The latter are more important for drought analysis.	DA
Cyclicality	The drought characteristic experiences fluctuations between low and high values.	DA
Persistence	The magnitude of the drought characteristic remains more or less unchanged for intervals of time.	DA
Hotspots	A time or place with considerably higher values than neighbours. Hotspots can be identified when the drought characteristic shows higher-than-average magnitudes at specific times or places.	DA, *nc*, *lc*
Cohesion	When drought areas tend to be together.	DA, *nc*, *lc*
Fragmentation	When the total spatial extent of the drought is composed of many clusters (fragments).	DA, *nc*, *lc*
Similarity	The drought characteristic shows similar patterns of occurrence.	DA, *nc*, *lc*
Dispersion	The drought characteristic presents a wide range of values.	DA, *nc*, *lc*
Trend	When the drought characteristic values experience a marked increase or decrease.	DA, *nc*, *lc*

5.3 SUMMARY AND CONCLUSIONS

The method entitled Spatio-Temporal ANalysis of Drought (STAND) was introduced in this chapter. The use of STAND is illustrated through different case studies; these results are presented in the following chapters. STAND toolbox is available at www.researchgate.net/project/STAND-Spatio-Temporal-ANalysis-of-Drought.

6

COMPARISON OF DROUGHT INDICATORS

6.1 INTRODUCTION

This chapter focuses on using daily spatially distributed hydro-meteorological data for calculating three types of drought: meteorological, agricultural and hydrological (Chapter 2.2). In this chapter, the performance of various drought indicators was examined to identify and characterise drought according to two catchment aggregation methods: catchment-aggregated time series of the drought indicator and the percentages of areas in drought. Historical droughts relevant to the study area were also compared in both methods. The case study is the La Sierra River basin, Mexico.

6.2 METHODS AND DATA

In a past study, Diaz Mercado et al. (2015) used the distributed hydrological model CEQUEAU-Idrisi to calculate runoff and evaporation for the La Sierra River basin, Mexico (see Chapter 4.1.2). In the present research, drought indicators were calculated for the following aggregation periods: 1, 3, 6, 9, 12 and 24 months. The following drought indicators were applied: the SPI (McKee et al., 1993), SPEI (Vicente-Serrano et al., 2010), SRI (Shukla and Wood, 2008), and the Standardized Evapotranspiration Deficit Index (SEDI) here introduced. Details on the description and methodology for calculating SPI, SPEI and SRI were presented in Chapter 2.2.

The performance of each indicator is examined to identify drought according to two catchment aggregation methods: (1) catchment-aggregated time series of the drought indicator and (2) the percentages of drought areas (PDAs) identified by the NCDA method (Corzo et al., 2011) (Chapter 5.2.2).

Standardized Evapotranspiration Deficit Index (SEDI)

SEDI is a drought index based on the evapotranspiration deficit (ED). ED can be calculated with potential evapotranspiration (Ep) and actual evaporation (E). Some formulations consider the difference between these two variables (Ep-E). Another way to calculate ED is by using Ep and E's relative difference with respect to Ep, i.e. (Ep-E)/Ep. In this last equation, ED goes from 0 to 1 with the lowest value being the worst condition.

For calculating drought with the use of ED, Diaz Mercado et al. (2016) proposed a drought index that follows the SPI methodology but using ED instead of P. ED is calculated as (Ep-E)/Ep. In this research, SEDI is calculated as proposed by Diaz Mercado et al. (2016) (Figure 6.1). For fitting ED, the empirical probability of Kaplan-Meier is followed (Sect. 2.2.2).

Theoretically, with the use of Ep, it is possible to compute the amount of water lost through evaporation or transpiration if there is an adequate water supply. Therefore, the inclusion of Ep in drought calculation is more desirable than the use of only P. However,

Ep captures the most extreme situation of the evaporation situation. Therefore, a closer evaluation of the condition on the ground is necessary. This limitation can be overcome with the use of *E* that can be calculated following remote sensing approaches or hydrological modelling, although empirical formulas can also be applied. By taking into account *E*, which is the volume of the water that is really evaporated directly from the soil, SEDI can capture a more accurate image of the ground's drought condition.

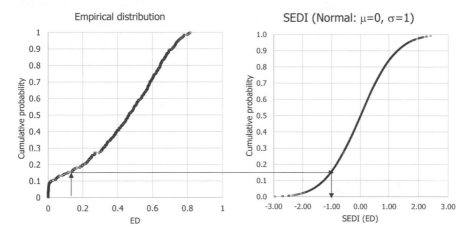

Figure 6.1 SEDI calculation (Diaz Mercado et al., 2016). First, evapotranspiration deficit (ED) is fitted by the empirical probability of Kaplan-Meier (left). Second, the cumulative probability of fitted ED is used to compute the SEDI value with consideration for normal distribution (right).

6.3 RESULTS AND DISCUSSION

6.3.1 First method: catchment-aggregated drought indicator

Figures 6.2a to 6.2f illustrate the time series of catchment -aggregated indicators at the time steps of 1, 3, 6, 9, 12 and 24 months, respectively. The dotted line at the value of -1 represents the threshold used to indicate a drought event. All indicators show similar patterns of wet and dry periods.

SPI-1 did not detect any drought between June and October, but SRI-1 detected that the basin was in drought in June 1994 (figure 6.3a and 6.3b). This drought event is consistent with a historical drought event reported in 1994.

SEDI-12 and SEDI-24 captured a prolonged groundwater drought period in the early 70s, but the other indicators failed to do so. This period is also consistent with a historic drought in 1969, in which water supply problems and dry wells were reported.

The SPI and SPEI presented a high correlation for the six-time steps: 1, 3, 6, 9, 12 and 24 months (Figure 6.4). The SPIs and SPEIs correlate with the SEDIs, except for the long-term steps of 12 and 24 months. The SRIs were found to correlate with their counterpart SPIs, SPEIs and SEDIs.

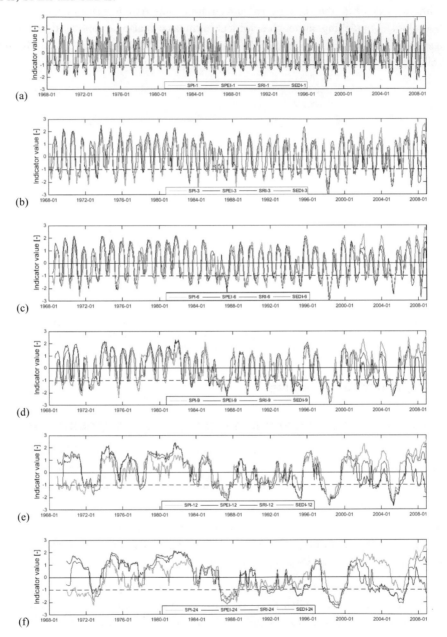

Figure 6.2 Time series of catchment-aggregated drought indicators for the time steps of 1, 3, 6, 9, 12 and 24 months.

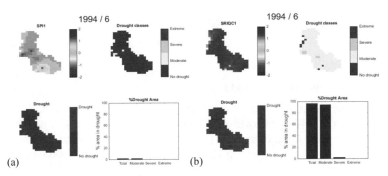

Figure 6.3 Comparison of the results of (a) SPI-1 and (b) SRI-1 for the selected month of June 1994.

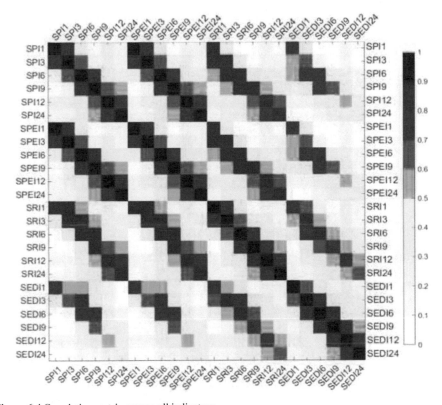

Figure 6.4 Correlation matrix among all indicators.

6.3.2 Second method: drought areas

Figure 6.5 illustrates the percentage of the basin area in drought (i.e. when the indicator value was ≤ -1). The matrices correspond to the indicators of the SPI, SPEI, SRI and SEDI, and to the time steps of 1, 3, 6, 9 and 12 months. The SPI, SPEI, SRI and SEDI followed roughly the same pattern for all time steps.

For the short-term steps of 1 and 3 months, the SRI and SEDI identified droughts in places where the SPI and SPEI failed to do so. The long-term steps of 6 and 9 months were used principally as indicators of hydrological drought; as shown by the SPI and SPEI, these drought events tend to be longer, which is in agreement with the literature.

The SRI and SEDI presented similar patterns of long periods of drought for the time steps of 6 and 9. These indicators also detected much longer periods, such as that observed in 1966 (Figure 6.5, SRI-9 and SEDI-9).

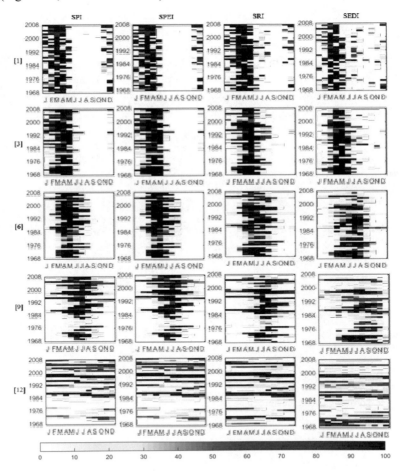

Figure 6.5 Percentage of drought area (PDA) for each drought indicator (columns) and each time steps (rows).

48

For the time step of 12 months, the SPI, SPEI and SRI followed roughly the same pattern. The SEDI, however, detected drought events in years where the others failed to do so. Around 1970, the SEDI identified a long drought period using the 12-month time step for detecting groundwater drought. According to historic drought data, 1969 was a year in which water supply problems and dry wells were reported. The SEDI was able to detect this drought even in the early 70s (Figure 6.2f).

Percentages of drought area (PDAs) of the SPI and SPEI lagged behind those of the SRI and SEDI; this was verified by correlation coefficient calculations of PAD time series per the SPI, SRI and SEDI (Figure 6.6). The SPEI was not considered for this because the indicator presented a similar pattern to that of the SPI; as such, the SPI results were extrapolated to the SPEI. For the time steps of 3, 6 and 9 months, the SPI's PAD time series exhibited a more significant correlation for a lag of 1 month. This result in the lag between SPI and the other drought indicators is expected because hydrological drought tends to start after the meteorological drought.

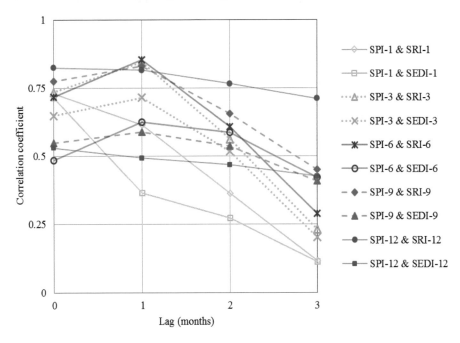

Figure 6.6 Correlation analysis between PDA time series of SPI - SRI, and SPI - SEDI.

49

6.4 SUMMARY AND CONCLUSIONS

These results demonstrate that meteorological drought indicators do not identify all drought events for the time steps of 1 and 3 months when considering drought areas. For the 3-, 6- and 9-month time steps, meteorological drought indicators tended to identify drought onset, but lagged behind the detections by hydrological drought indicators. For the long-term time steps of 12 and 24 months, the results indicate that agricultural and hydrological droughts indicators are more appropriate than meteorological ones. Further, drought in a catchment is best monitored via joint evaluation that combines meteorological drought indicators with hydrological and agricultural ones. This allows drought events and their spatio-temporal evolution to be identified with greater accuracy. The results also demonstrate that spatially distributed runoff and evaporation simulations can improve drought identification because the time series of catchment-aggragted drought indicators in some cases presents similar results giving the wrong idea that two indicators behave in a similar way, when their spatial distribution is totally different.

Comparisons of drought indicators and their types can benefit greatly from spatio-temporal drought analysis (second method). Considering the spatial extent of droughts allows the analysis of patterns such as seasonality, hotspots and lags between drought events to be calculated by different drought indicators. This last aspect is essential for forecasting matters.

7

MACHINE-LEARNING APPROACH TO CROP YIELD PREDICTION

7.1 INTRODUCTION

Drought continually hits many regions across the world. It negatively affects (impacts) various human activities such as agriculture, which not only generates economic losses but can also trigger famine, causing millions of deaths (Below et al., 2007; Food and Agriculture Organization of the United Nations (FAO), 2017; Kim et al., 2019; Sheffield and Wood, 2011; World Meteorological Organization (WMO), 2006). Hence, methods that help to improve strategies for drought mitigation are necessary. Within these methods are those that allow predicting the impacts of drought.

Assessments of drought impacts confirm that the presence of drought on human activities can be devastating. For instance, the Food and Agriculture Organization of the United Nations (FAO) calculated the damage and losses in the agricultural sector caused by five types of hazards, including drought. FAO estimates that drought causes damages and losses to this sector by up to 80% (FAO, 2017). Although multiple factors are involved in agriculture, drought often plays the primary role, as literature confirms (Dai, 2011; FAO, 2017; Kim et al., 2019).

The assessment of drought impacts on agriculture can be performed in terms of crop yield. FAO defines crop yield as the measure of the yield of a crop per unit area of land cultivation (in kg/ha or ton/ha) (FAO and DWFI, 2015). For the assessment of crop yield under drought affectation, physical models based on crop properties turn out to be more comprehensive and descriptive (Reynolds et al., 2000; White et al., 1997; Wu et al., 2016). However, an important barrier to such models' realisation is the lack of detailed crop data and the difficulty representing all the processes involved in all stages of crop development (Reynolds et al., 2000; Wu et al., 2016).

Statistical and machine-learning (ML) models, which involve mathematical equations to calculate the output of a model with suitable input(s), can be used to assess crop yield impact by drought without considering any biological or physical process of the crop but the analysis of the input and output data (Chlingaryan et al., 2018; Rahmati et al., 2020; Udmale et al., 2020; van Klompenburg et al., 2020). There have been studies where various inputs, ML techniques, and architectures (configurations) have been tested for crop yield prediction (e.g., Chlingaryan et al., 2018; van Klompenburg et al., 2020). However, the spatial extent of drought (area) is an input that has not been fully explored previously to crop yield prediction. The prediction refers to the calculation of crop yield at the end of the growing season (harvesting) with information available before or during the crop development season (pre-harvesting).

This chapter introduces an ML approach to calculate seasonal crop yield (CY) with the monthly percentage of drought areas (PDAs) as input. The ML approach comprises two components. Each component includes a set of the following types of ML models: polynomial regression (PR) and artificial neural network (ANN). The goal is to compare both ML models (ANN and MR) and use them as an integrated tool to support the decisions made based on the crop yield prediction. The logic is as follows. PR provides the prediction where the crop yield calculation is "clear" to the performer (the end user) because she/he has access to the equations that have a straightforward interpretation. For its part, ANN is used as the most accurate model, although the calculation of the output is not as "clear" as in the case of PR due to the difficulty of interpreting the structure of the resulting ANN.

Three East Indian states where agriculture plays an important role were chosen as a case study. These states are Bihar, West Bengal, and Odisha. ML models were built for the period 1967-2015. ML models aim to predict rice crop yield since rice is the most cultivated crop in the region. The ML approach was applied separately to the three regions (states).

Crop yield prediction in India

In India, as in many other countries, the official crop yield prediction is mainly based on conventional data collections techniques such as ground-field visits (Reynolds et al., 2000; Sawasawa, 2003). The crop yield is measured through crop cutting experiments carried out over sample crop areas. In this country, principal crops' calculations of area and yield are released through the Directorate of Economics and Statistics, Ministry of Agriculture (DESMOA). The production (in kg or ton) of a specific crop is calculated by multiplying the whole field area by its crop yield. The crop production is needed for the decision-makers to take various policy decisions relating to pricing, marketing, distribution, exportation and importation.

The Kharif season, as it is locally known, represents about 80% of the annual rainfall (Naresh Kumar et al., 2012). This monsoon season generally goes from June to October. In this season, the highest agricultural production is obtained. Estimation of Kharif crop yield and production is released four times during the year with different levels of sophistication and precision, where the last one is considered the most accurate. The first calculation is presented in September, the second one in January, the third one in March/April, and the fourth, and the last one in June/July. It should be noted that the last two calculations of crop yield and production become available much after the crops have already been harvested in December/January. From the four calculations, the first two can be considered as predictions. These two first predictions serve as primary estimations about how much the final yield and production will be.

The existing ground-field visits-based agricultural forecasting system provides reliable information; however, it lacks pre-harvesting forecasting. This limitation motivated the creation of a new satellite-based forecasting system to have information at the early stages of crop growth. This system is called the National Crop Forecasting Centre (NCFC) (Sawasawa, 2003). NCFC is continuously verified and continuously updated. Although

53

NCFC advances the one based on ground-field visits, data needed for its execution could be not always available. Therefore, it is necessary to explore other solutions. In this study, it is not intended to replace the previous and new forecasting systems, but to provide a complement to corroborate both estimates, and in a broader sense, to provide the scientific community with an approach to crop yield prediction with information on the spatial extent of drought.

7.2 ML MODELLING METHODOLOGY

The experiment was carried out with the following methodology that involves the ML construction. The next paragraphs show each step in detail. These steps are (1) data preparation, (2) input variable selection, (3) polynomial regression models calculation, (4) artificial neural network models calculation, and (5) models application and combination.

7.2.1 Step 1. Data preparation

Two types of data were prepared, the time series of crop yield and the percentage of drought areas. For data preparation, three tasks were carried out (1) data retrieving, (2) drought areas calculation, and (3) data de-trending.

Data retrieving

Chapter 4.2.2 shows what corresponds to data retrieving for crop yield (CY) and the drought indicator. CY data correspond to the largest growing season. CY time series has a value for each year for the period 1966-2015 (49 years). On the other hand, drought indicator data is on a monthly basis for the period 1901-2015. The spatial resolution is half a degree.

Drought areas calculation

The drought areas were calculated following the methodology presented in Chapter 5.2.2. The monthly time series of drought areas were calculated for three regions (states): Bihar, Odisha, and West Bengal. The mask in raster format was built for each region. This mask is an array of ones and zeros, where the value of 1 indicates the land. The number of cells in each mask is 63, 54, and 31 for Bihar, Odisha, and West Bengal.

Upon SPEI data, the cells in drought were identified at each time step. The threshold $T =$ -1 was used to calculate cells in droughts. This threshold is widely used to identify a cell in drought when working with standardised indices such as SPEI. Then, the time series of percentage of drought areas (PDAs) were calculated for each SPEI dataset of 1, 3, 6, 9, and 12 months of aggregation period. PDAs' time series are indicated as PDA1, PDA3, PDA6, PDA9, and PDA12, respectively.

Data de-trending

In time series applications, data stationarity is typically assumed when modelling. However, the present study uses crop yield as its input variable, which is non-stationary in nature. The crop yield depends on factors that affect its trend, such as drought, flood, cultivars and its own management. Therefore, it is advisable to remove short-term fluctuations in crop yield before constructing the model (Montesino Pouzols and Lendasse, 2010).

Among the methods available to de-trend data, the 'first difference' method is popular due to its simplicity. In this method, the trend is removed from the time series by subtracting the previous value $x^*(t-1)$ from the current one $x^*(t)$, as shown in Eq. 7.1. The de-trended value for the first time step ($t=1$) is not calculated. The length of the de-trended time series is $n=m-1$, where m is the length of the original time series. The de-trended data $x(t)$ has the same units as the original data $x^*(t)$.

$$x(t) = x^*(t) - x^*(t-1)$$ (Eq. 7.1)

After obtaining the CY and PDAs time series, the trend of each was removed with Eq. 7.1. For the case of CY, the de-trended time series retained one value per year. As noted, the method for removing the trend does not generate the value for the first time step; therefore, the de-trended CY data corresponds to the period 1967-2015.

In the case of PDAs, Eq. 7.1 was applied as follows. Because the PDA data is monthly, i.e. 12 values per year, and CY data is seasonal, i.e. one value per year, PDA time series were extracted for each month. The monthly values for January were extracted for each year and so on until December. These twelve time series were compiled for each of the five PDA1, 3, 6, 9 and 12 time series. A total of 60 time series (12×5) were obtained. To refer to these time series, a number was added to indicate the month. In this way, for example, the time series PDA3_7 indicates the percentages of drought area for July calculated from SPEI3. Eq. 7.1 for the removal of the trend was applied to each of the 60 time series. All SPEI databases run from 1901-2015. For the construction of the ML models, the common period 1967-2015 was chosen.

7.2.2 Step 2. Input variable selection

In an ML model, the input, known as the predictor, is generally made up of independent variables. Often these variables are arranged in different ways to determine the best model input representation. An example arrangement is the selection of the independent variable using different previous time steps, such as $t-1$ (the previous time), $t-2$ and so on. When using drought indicators as the predictors, these arrangements include the different aggregation periods (i.e. different aggregation periods are tested). The idea is not to include all the variables and all their different possible arrangements but rather to find the best ones and discard those that do not contribute significantly to the model's results. Other arrangements of the input variable include the average, or other statistics, over a period.

There are different methods for selecting input variables. Based on the procedure, these methods are classified into model-based and filter types (May et al., 2011). The first includes all those where the model runs, and based on its performance, a specific variable is chosen or discarded. The latter include methods where the variable is chosen *a priori* through a generally statistical process and does not require the model to be run. Correlation analysis, which falls under the second category, is often chosen for its simplicity and wide application. Correlation is calculated between the time series of the output variable (CY in this case) and the different input variables, including the various arrangements.

In this study, for the selection of the relevant input variables, the correlation analysis was done. The correlation was calculated between the de-trended time series of the seasonal CY and the 60 PDAs. As mentioned before, due to PDAs are monthly and CY is seasonal, 12 time series of PDAs were prepared, one per month, for each aggregation period. The PDAs were then correlated with the CY. Another option could be to use an average value of monthly PDAs, such as the average of the PDAs of the months of the cultivation period, or something similar. However, we opted to identify the PDAs of the months that have the highest correlation with the seasonal CY.

The approach of the selection of the most correlated PDAs was chosen for two main reasons. On the one hand, rice responds to the climate variations differently from one growth stage to another over the year, which could be better captured with the information of some months than others. On the other hand, different types of drought (i.e. meteorological, agricultural, and hydrological) are expected to affect (impact) the crop yield to different degrees. This level of affectation could be taken into account either by using different hydrological variables or selecting different aggregation periods of the meteorological variables, as in this case. An average of PDAs of the same or different aggregation periods could "hide" a significant drought area that could contribute more (or less) to the final crop yield. In addition, in this research, ML models are built to be used at different stages of crop cultivation, i.e. models to be applied in June, July, and so on, each of them with a different expected degree of accuracy. Therefore, the use of time series for each month extracted from the PDAs for all the different aggregation periods is more appropriate.

Based on the correlation coefficient, the input variables were selected. In total, 15 sets of input variables (Table 7.2) were selected for each month from January to December. Each set is made up of different PDA time series, out of the 60 de-trended PDAs. The number of variables is different in each set. These sets of input variables are presented in the results section. All sets include the de-trended CY from the previous year (CY_{t-1}). CY_{t-1}

was used because, in the particular case of the study area, CY of the current year is planned to be reached based on data of the previous year.

7.2.3 Step 3. Polynomial regression models calculation

For the case of PR, four types of model were tested (Table 7.1). All the PR models were built for each month from January to December following Eqs. 7.3 to 7.6. For each month, different 15 sets of combinations of input variables were tested in each PR model. The best PR model was identified for each month following the RMSE criterion (Eq. 7.7). MATLAB software was used for implementation.

PR is an extension of linear regression that allows the use of more than one input variable to calculate the output variable. PR is expressed with Eq. 7.2.

$$y = b_0 + \sum_{i-1}^{n} b_i x_i + e \qquad \text{(Eq. 7.2)}$$

In Eq. 7.2, y is the output variable, also known as the response, which in this case is the crop yield. The term x_i is the i-th input variable (predictor) from a total of n variables. The regression coefficients vector is represented by b. From the coefficients vector, b_0 is known as the intercept. The vector of errors is indicated by e.

Table 7.1 shows four formulations of PR. The PR models are indicated as linear, pure-quadratic, quadratic and interactions. Descriptions of each and their equations are presented in Table 7.1 (Eqs. 7.3 to 7.6). The input variable (x_i) is selected based on the correlation analysis.

Table 7.1 Polynomial regression (PR) types followed in this study.

PR type	Equation	Description
Linear	(Eq. 7.3) $y = b_0 + \sum_{i=1}^{n} b_i x_i$	It has an intercept and linear terms of predictors
Pure-quadratic	(Eq. 7.4) $y = b_0 + \sum_{i=1}^{n} b_i x_i + \sum_{i=1}^{n} b_{n+i} x_i^2$	It has an intercept, as well as linear and squared terms of predictors
Quadratic	(Eq. 7.5) $y = b_0 + \sum_{i=1}^{n} b_i x_i + \sum_{i=1}^{n} b_{n+i} x_i^2 + \sum_{i=1}^{n-1} \sum_{j=i+1}^{n} b_{2n+(i-1)n-\frac{(i-1)i}{2}+(j-i)} x_i x_j$	It has an intercept, linear and squared terms and all products of pairs of distinct predictors
Interactions	(Eq. 7.6) $y = b_0 + \sum_{i=1}^{n} b_i x_i + \sum_{i=1}^{n-1} \sum_{j=i+1}^{n} b_{n+(i-1)n-\frac{(i-1)i}{2}+(j-i)} x_i x_j$	It has an intercept, linear terms of predictors, all products of pairs of distinct predictors and no squared terms

57

The best PR model is identified from four types using the root mean square error (RMSE) criterion. The RMSE is calculated between the observations (o) and the predictions (p), as shown in Eq. 7.7.

$$RMSE = \sqrt{\frac{\sum_{i=1}^{n}\left(o_i - p_i\right)^2}{n}}$$
(Eq. 7.7)

RMSE is one of the most widely used criteria in the comparison of observations and model calculations.

7.2.4 Step 4. Artificial neural network models calculation

ANN is a method loosely based on imitating the basic functionality of neurons (i.e. the working units of the human brain) (Govindaraju, 2000; Maier and Dandy, 2000). The input variables (predictors) are connected to each other through mathematical formulations that allow complex non-linear relationships to be represented. These connexions are symbolised as nodes interconnected within a network aimed at calculating the output variable (response).

Of the different proposed ANN architectures (network designs), one of the most widely used is the feedforward neural network (FFNN). The FFNN is schematised by a series of nodes located in one of three layers: input, hidden or output. The number of input nodes is equal to the number of input variables in the input layer (Elshorbagy et al., 2010). This first layer is in turn connected to the hidden layer, which receives this name because the connections made there may not be immediately evident to the model performer. In this hidden layer, the number of nodes is not defined by default; rather, the greater the number of nodes, the more complex the model. Finally, the nodes of the hidden layer are connected to those of the output layer. In a single-output variable problem, there is only one node. ANNs are typically trained by non-linear optimization gradient-based algorithms, e.g. the Levenberg-Marquardt algorithm.

In the ANN setup, the number of nodes of the input layer was equal to the number of variables of the respective combination. The number of nodes in the output layer was one and corresponded to the seasonal crop production (CY). An iteration optimization procedure was carried out regarding the hidden layer, varying the number of nodes from 1 to 10. For each number of nodes, 100 iterations were done, being 1,000 in total. For reproducibility of the results, the random values were set to default at the beginning of the change of number of nodes. For each month, from January to December, the ANNs were built. MATLAB software was used to implement the ANNs with the Levenberg-Marquardt algorithm for training. In each of the ANNs, 85 % of the data was used for training-validation, and the rest for testing (verification). The best model corresponding to each number of hidden nodes was identified, i.e. ten models per month and the best model for each month. RMSE was used to identify the best models. RMSE was calculated for (1) the training-validation dataset (RMSE_cal), (2) the testing dataset (RMSE_test), and (3) the entire period (RMSE). In all the cases, the final (best) model was chosen based on RMSE for the entire period.

7.2.5 Step 5. Models application and combination

Once the best ML models, PR and ANN, are known, the pair of models were selected for each month. Depending on the performance of these models (and experience of their use), they can be used either separately or combined, e.g. being run in parallel so that a modeller could see the cases when models produce different results. An alternative is to use a dynamic weighting of the models' outputs (e.g. with the weights being proportional to the historical performance) to form a "model committee".

7.3 RESULTS AND DISCUSSION

7.3.1 Correlation analysis

Figures 7.1 (a, b, and c) shows the correlation between the de-trended crop yield and the percentage of drought areas.

It is observed that the correlation is different over the year in the three regions. In general, in all cases, the correlation coefficient increases until it reaches a maximum and then decreases. The month in which the maximum value is reached is different for each region. For Bihar, it is in July. For West Bengal, there are four months with this pattern, June, July, October, and November. Finally, for Odisha, it is from October to December.

These results can be useful for monitoring agricultural drought. For example, in Bihar, the drought areas calculated from SPEI6 (PDA6) show a maximum correlation in July. This correlation value means that the previous six months' accumulated effect is crucial for the crop yield of the Kharif season, which covers more or less from June to November. In general, for Bihar, results similar to PDA6 are observed for PDA3, 9 and 12. For West Bengal, something similar pattern happens concerning the peaks, except that there are two, one corresponding to PDA1 and 3, and the other for PDA6, 9, and 12. The first peak of PDA1 and PDA3 may indicate that it is crucial to pay attention to the immediate period conditions of one to three months. In the case of the second peak, the medium and long-term conditions, 6 to 12 months, are more important to monitor for the harvest month. In the case of Odisha, the peak occurs until the end of the growing season, almost for all periods of aggregation. Hence, the condition before the growing season is decisive for the crop yield. Figure 7.1 (d) shows the percentage of agriculture irrigated and rain-fed in the three regions. For Bihar and West Bengal, about half is by irrigation, while in Odisha, only 35%. Perhaps this percentage for Odisha explains why the correlation coefficients for Odisha are higher than for Bihar and West Bengal. Odisha is more dependent on rain for agriculture. This condition is best captured using drought areas calculated with SPEI that considers precipitation and evaporation for its calculation.

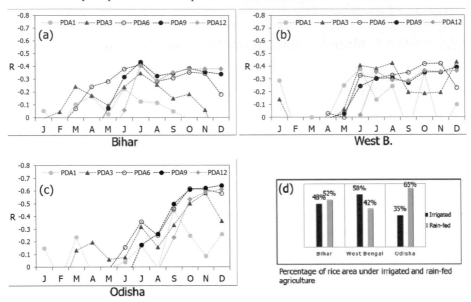

Figure 7.1 Correlation between de-trended crop yield and percentage of drought areas at different aggregation periods for Bihar (a), West Bengal (b), and Odisha (c). Percentage of rice area under irrigated and rein-fed agriculture (d).

The following pattern is observed in the three cases in Figures 7.1 (a, b, and c). In general, the correlation coefficients between CY and PDAs increase according to the aggregation times, i.e. PDA1, PDA3 has a better correlation for the first months of the year, then PDA6 begins and PDA9 and 12 successively.

Each respective PDA time series reaches a maximum (or maximums) of correlation, and then correlation decreases. According to this pattern, the 15 combinations of input variables shown in Table 7.2 were selected. As previously mentioned, the CY of the previous year was included in all combinations, and this time series is indicated as CY_{t-1}. Combinations 1 to 5 only include a PDA time series. Combinations 6 to 9 are PDA pairs that were calculated with SPEI of successive aggregation times. For example, combination 6 forms PDA1 and 3, combination 7 includes PDA3 and 6, and so on. Similarly, combinations 10 to 13 are proposed, but for triples. Combinations 13 and 14 are fourfold. Finally, the last combination (15th) is made up of all the PDA series.

As mentioned in the previous section (Sect. 7.2), the models were built for each month using the 15 combinations corresponding to each month. For example, for January, the monthly series of PDAs extracted for January were used, these are PDA1_1, PDA3_1,

PDA6_1, PDA9_1 and PDA12_1. Then, PDA1_1 to PDA12_1 were used following the 15 combinations.

Table 7.2 Input sets (combinations) for ML models.

Input set (combination)	Input variables
1	CY_{t-1}, PDA1
2	CY_{t-1}, PDA3
3	CY_{t-1}, PDA6
4	CY_{t-1}, PDA9
5	CY_{t-1}, PDA12
6	CY_{t-1}, PDA1,3
7	CY_{t-1}, PDA3,6
8	CY_{t-1}, PDA6,9
9	CY_{t-1}, PDA9,12
10	CY_{t-1}, PDA1,3,6
11	CY_{t-1}, PDA3,6,9
12	CY_{t-1}, PDA6,9,12
13	CY_{t-1}, PDA1,3,6,9
14	CY_{t-1}, PDA3,6,9,12
15	CY_{t-1}, PDA1,3,6,9,12

7.3.2 Polynomial regression (PR) models

Some examples of the polynomial regression models' results are presented in the figures below for all three regions, Bihar, West Bengal, and Odisha. The data was divided into training-validating and testing dataset. PR model result of Bihar region is shown in Figure 7.2. RMSE shows that this model has less accuracy than Bihar's ANN model, as was expected. The model was unable to predict all the unseen data successfully due to simplicity, therefore. For the Odisha region (Figure 7.4) also RMSE is lower than the ANN. The yield of this state considerably fluctuates over time.

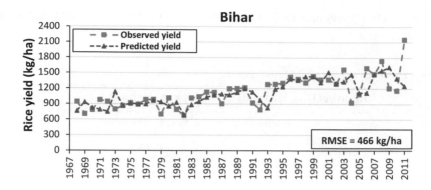

Figure 7.2 Polynomial regression model for Bihar.

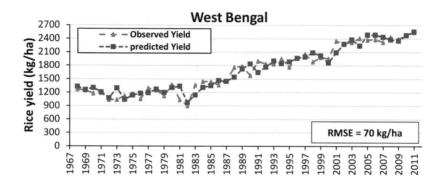

Figure 7.3 Polynomial regression model for West Bengal.

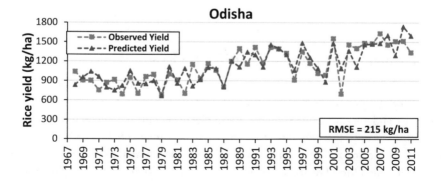

Figure 7.4 Polynomial regression model for Odisha.

7.3.3 Artificial neural network (ANN) models

The figures below show the example of the results of the trained and tested neural network models for each region. For Bihar, in Figure 7.5, it is clear that the model, during the training, successfully captured the drops or reduction in rice yield. RMSE for the tested model is relatively high (311 kg/ha) than the state average from 1967 to 2015 (1140 Kg/ha). This region's error is about 27%, which can be attributed to data restriction, where more data is needed to train the model. From the previous figures, the results of the regional rice yield prediction models were presented. It is evident that ANN models' perform better than MR models for all three regions. In addition, in the West Bengal region (Figure 7.6) ANN model has high accuracy (RMSE = 62 Kg/ha), which accounts only for 3.6 % of the average yield over 1967 to 2015. Odisha region prediction accuracy indicated an RMSE relatively low (59 Kg/ha) when compared to the state average (1130 Kg/ha), which is about 5.2 % (Figure 7.7). In the case of Odisha, the model was trained and tested well; almost all fluctuations were captured. It is important to note that only four last years have been used to test the model, and the other previous four years used for validation.

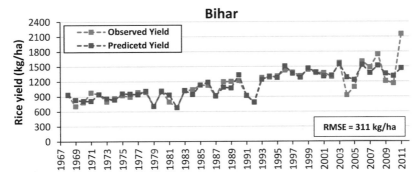

Figure 7.5 ANN model for Bihar.

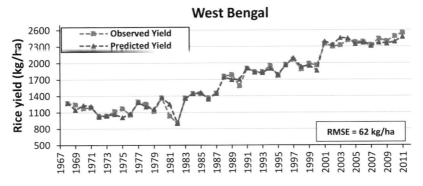

Figure 7.6 ANN for West Bengal.

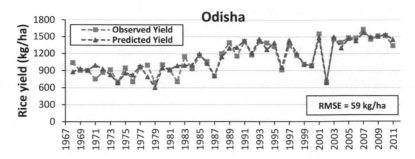

Figure 7.7 ANN for Odisha.

7.3.4 ML modelling limitations

The limitations of the presented approach are the following.

(1) To determine drought areas, a threshold value of the Standardised Precipitation Evapotranspiration Index (SPEI) drought index (SPEI ≤ -1) was used. Using one value might lead to over or underestimation of the actual drought impacts over crop yield.

(2) Gridded data of SPEI at spatial resolution (0.5°X 0.5°) was used in this study over each region individually. Using such a coarse spatial resolution on different region sizes does not capture the drought's reality on the ground, leading to over or underestimating the drought area.

(3) The study area has a diverse ecosystem of irrigated and rain-fed land, which may influence the correlation between PDA and crop yield more or less.

(4) This study assumes that drought is the only causative factor; however, floods negatively impact crop yield in the region, thus in the total production in the regions. Flood impacts are not considered in the models.

(5) Many other factors might influence rice yield, such as market, technologies, management, etc. In this study, it was assumed that drought plays the main role.

(6) Insufficient crop yield data for the ML model building was an issue because the CY time series only had one value for each year.

7.4 SUMMARY AND CONCLUSIONS

This chapter introduced an ML approach for predicting seasonal crop yield (CY) with a monthly percentage of drought areas (PDAs) as input. The ML approach comprises two components. Each component employs two ML models: polynomial regression (PR), and artificial neural network (ANN). The goal was to compete the ML models (ANN and PR) with each other and use them as an integrated tool to crop yield prediction.

The following conclusions are drawn from this study.

- Based on the RMSE of ML and ANN models, results show drought area to be a suitable variable to predict crop yield.

- The correlation analysis between PDA and CY showed high negative correlations in Odisha. The correlation gradually decreases in Bihar and West Bengal. These correlation values can be because West Bengal has better access to irrigation facilities than Odisha and Bihar.

- On comparing ANN models and PR models, the ANN were more accurate than PR models to predict crop yield for all regions. This could have been expected since the drought–crop relationship is a highly non-linear problem.

- It can be concluded that neural network has a high capability to predict crop yield in the pre-harvesting stage with good accuracy, considering SPEI which uses climate variables such as precipitation and temperature (for evapotranspiration calculation).

From the analysis and findings of this research, the following recommendations can be provided for further improvement.

- Sensitivity analysis should be performed to identify the parameters that can impact the model results. For instance, different spatial resolutions of SPEI and different SPEI thresholds should be investigated to determine their impact on the model results.

- Wet extreme events should be considered, especially in the flood-prone regions such as the Coastal areas of West Bengal and Odisha region and North Bihar where floods also influence crop yield.

- Non-climatic factors such as econometric, fertilizers, and management practices are considered because they influence crop yield.

- In order to improve the model accuracy, more input data should be used in further studies. For crop yield, this can be estimated by remote sensing techniques on a monthly basis so that the ML models can be built for this temporal resolution.

- The performance of other ML models has to be investigated, especially committee (ensemble) methods like random forests or boosting methods. In the case of data at scales less than monthly, the use of deep learning algorithms (e.g. LSTM networks) could be recommended to explore.

8
VISUAL APPROACHES TO DROUGHT ANALYSIS

8.1 INTRODUCTION

Drought is a natural phenomenon whose effects (impacts) can cause numerous deaths and considerable economic losses (Below et al., 2007; Sheffield and Wood, 2011; Wilhite, 2000). On the one hand, its analysis is essential for the understanding of its drivers and development. On the other hand, its monitoring is vital for quantifying its intensity and assessing its possible impacts (World Meteorological Organisation (WMO), 2006). During the performance of both drought analysis and monitoring, the following tasks are often carried out. (1) Calculation and comparison of the drought characteristics (features), such as spatial extent (area), duration, and intensity. (2) Identification of locations or periods with higher-than-the-average values of a given drought characteristic. (3) Analysis of drought patterns, i.e. how a given drought characteristic is presented over time. The information generated during or as results of the previous tasks is referred to here as drought-related data.

In the tasks mentioned above, some data visualisation are frequently used to explore or explain drought-related data. To analyse or show how a drought characteristic change over time, line graph or area chart are often selected, while to analyse the correlation between different drought characteristics, scatterplots are preferred. Histograms are often used to present the statistical distribution, whereas boxplots are picked up to visually summarise the statistics (e.g. mean, standard deviation, maximum, minimum). To identify patterns on the variation of drought over time, heat maps, also referred to as colour-coded table, are considered. In this columns-rows arrangement, the information is organised to show how a given drought characteristic changes over time. The colour assigned to each cell relates to the magnitude of the analysed drought characteristic. The adjacent cells with similar colour aligned in columns (or rows) allow identifying periods with high (or low) intensity.

Data visualisations mentioned above constitute a useful catalogue to explore and explain many aspects of the drought. However, the visual identification of spatio-temporal patterns in the development of drought is not always straightforward to perform with the existing catalogue. In particular, drought patterns such as periodicity, seasonality, among others are challenging to detect.

This chapter introduces approaches based on radial and polar charts to interpret and analyse the spatio-temporal variability of drought. These charts can be used to perform a descriptive analysis of drought frequency, intensity and trend. It is foreseen that they aid in comparing spatio-temporal patters of drought calculated with different water cycle components, such as precipitation-based vs soil-moisture-based. The visualisations also can be used for comparing the results of different drought calculation approaches. Another expected application is their use in drought propagation research, i.e. how drought moves throughout the terrestrial hydrological cycle coming from precipitation to runoff (Peters et al., 2003; Van Loon and Van Lanen, 2012).

8.2 METHODS AND DATA

8.2.1 Visual approaches to drought analysis

Graphs introduced in this chapter are designed to visually analyse the variation of drought characteristics calculated with the methodology presented in Chapter 5. PDA, number of events (clusters), and the average distance among clusters (lc) are the inputs to these diagrams. Consider the monthly PDAs of five years, shown in figure 8.1. In this series, percentages increase progressively by 1 % since January (J) of the first year. December (D) of the fifth year has a value of 60 %. In figures 8.2 to 8.4, the five-year PDA values are used to explain the elements that integrate the next three diagrams: (1) Polar Area Diagram (PAD), (2) AnnUal RAdar chart (AURA) and (3) MOnthly Spider ChArt (MOSCA).

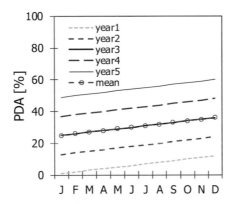

Figure 8.1 Example of the monthly percentage of drought areas (PDA) values of five years. PDAs are used to illustrate elements of diagrams shown in figures 8.2 to 8.4.

Polar Area Diagram (PAD)

Polar area diagram (PAD) is the polar chart that displays drought areas in a radial heat map layout. Drought areas are plotted in circular arrays (rings) following a calendar year. Colour of each polar area (segment), indicates the magnitude of the variable. PAD is inspired in the heat map, already used in drought analysis. PAD has been thought not to cut the cycle of data series by rows/columns as in heat maps, but to make it continuous by means of rings. This layout can facilitate the visual identification of drought duration and occurrence and the most severe years in drought. The darkest colours indicate the

69

highest values. Adjacent coloured segments represent extended periods of droughts. Episodes with few or no droughts are identified by voids (whitish segments).

When assessing drought development on various water cycle components (e.g., precipitation, runoff, and soil moisture), respective PDAs can be plotted in different PAD charts and by visual examination identify the drought variability and propagation. Comparing different drought proxies' datasets (e.g., SPI-/SPEI-based PAD) can also be done using different PAD charts.

Drought seasonality, i.e., repetitive and regular drought fluctuations over time, is observed in PAD by (quasi-) cyclical patterns. Pie-slice-shaped sections help estimate the beginning and end of such seasonal variations. How data on drought areas is organised in PAD aids in the identification of differences between drought patterns of different years (inter-annual drought variability) and drought changes within the same year (intra-annual drought variability).

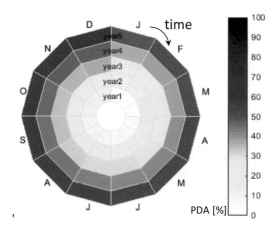

Figure 8.2 Polar Area Diagram (PAD): monthly percentage of the drought areas (PDAs) of the example shown in figure 8.1. PDAs are presented clockwise from January (J) to December (D). The darkest colour specifies the most severe drought.

AnnUal Radar chart (AURA)

Area charts and histograms often make use of colours to emphasise the highest values. A disadvantage of these charts is that when the data series is long, it is difficult to present it on the same horizontal axis (e.g., plotting a century of data). AURA aids to tackle this problem by drawing the PDA information in a circle, using a sequence of equiangular rays (radii) proportional to the PDA values (Figure 8.3). AURA helps to detect trends. Another advantage of this layout is that it shows outliers and similar values within the PDA series, which is useful when locating severe and similar drought periods. By its design, in AURA various spatial features of drought can be displayed. For instance, by

70

applying different colours to the radii, the number of clusters (*nc*) can be indicated. It is also possible to use more than one axis to show different drought features. Graphically, on different water-related variables, the spatio-temporal drought variability can be examined through different AURA charts. As well, when contrasting diverse drought proxies' datasets (e.g., SPI-/SPEI-based PAD).

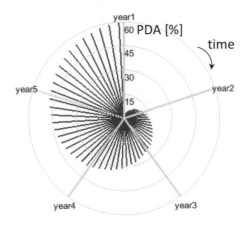

Figure 8.3 AnnUal RAdar chart (AURA): monthly percentage of drought areas (PDAs) of the example presented in figure 8.1. Data is displayed clockwise since the first year. Lengths of radii are proportional to PDAs and angles to the number of time steps.

Monthly Spider ChArt (MOSCA)

MOSCA is a kind of spider chart which allows analysing the variation of drought between years and months. In this chart, the time steps corresponding to one year period are represented in each segment. The length of the radius is proportional to the PDA magnitude (Figure 8.4). With the use of different colours, MOSCA enables visual comparison of distributions of PDAs between different years. This setup also permits to detect monthly outliers and (dis)similar drought distributions. Long-term stats of the PDAs, which provide a useful summary of the average and variation of drought's spatial extent, can also be depicted in this chart. Because PDA values are expressed in percentage, the coefficient of variation (CV [%]), i.e. the ratio of standard deviation and mean, is suitable for measuring long-term data variation. The box-plot chart inspired one proposed MOSCA variation: the box-plot MOSCA. In this chart, the PDA time series is shown as follows. Data distribution is displayed in the pie-slice-shaped sections. Each section refers to one month from January to December (J to D). Each box section's central line represents the median (50th percentile, q2); and the edges are the 25th and 75th percentiles (q1 and q3, respectively). The lower (w1) and upper (w2) whisker positions are defined by the Eq. 8.1 and 8.2, respectively. Three additional settings are included. First, PDA values are plotted to display data distribution using dots. Second, years with the most severe monthly droughts, those with the highest PDA values, are depicted with text tags. Third, (3) in the centre of the chart, a selected number of years with high monthly PDA values are displayed with text tags. These three settings can help identify years with severe droughts. MOSCA chart can also help analyse drought variability and propagation on different water-related variables by using different graphs or by plotting the annual periods of the variables in the same chart.

$$w1 = q1 - 1.5(q3 - q1) \qquad \text{(Eq. 8.1)}$$

$$w2 = q3 + 1.5(q3 - q1) \qquad \text{(Eq. 8.2)}$$

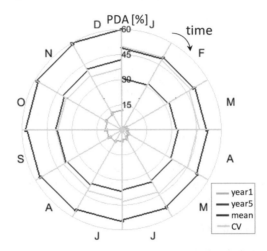

Figure 8.4 MOnthly Spider ChArt (MOSCA): First and fifth year of the monthly percentage of drought area (PDA) values of the example in figure 8.1. Annual periods are presented clockwise from January (J) to December (D). Radii are proportional to PDAs and angles to months. Coefficient of variation (CV) is in percentage.

8.2.2 Experiment setup

The visual approaches were applied in a case study to exemplify the radial and polar diagrams' use. Mexico was selected to apply the diagrams because of the frequent drought impacts presented over the country (Bhattacharya et al., 2014; CENAPRED, 2007; Florescano et al., 1980; Velasco, 2012). A description of the study area and the data used is presented in Chapter 4. Visual approaches were used for spatio-temporal analysis of agricultural drought variations in the 20th and the early 21st century (period 1901-2013). Drought was calculated with the Self-calibrating Palmer Drought Severity Index (scPDSI), provided in Chapter 4. The experiment was carried out in three sub-regions of different climate according to Köppen-Geiger climate classification (Kottek et al., 2006). Mexico was divided into three main climatic regions: equatorial, warm-temperature, and arid climates (figure 8.5). Masks were created to conduct the spatio-temporal analysis over the entire region and the three sub-regions. The spatial resolution of the four masks is 0.5 degree, which is the same that for the drought indicator database (Table 8.1).

Table 8.1. Summary of experiment.

Analysis period (no. years)	Drought proxy database	Radial diagrams	Masks	
1901-2013 (113)	scPDSI.cru	• PAD • Box-plot MOSCA • AURA	1. 2. 3. 4.	Country Arid Equatorial Warm-temperature

Visual analysis was carried out for a number of variables. The methodology of spatio-temporal analysis presented in Chapter 5 was applied here for the scPDSI drought index. From NCDA, the PDA was calculated, and from CDA, the number of clusters (*nc*) and the average distance among clusters (*lc*) were calculated as well. PDA time series were used to build PAD and MOSCA charts. PDA, *nc* and *lc* distances were employed to create AURA charts.

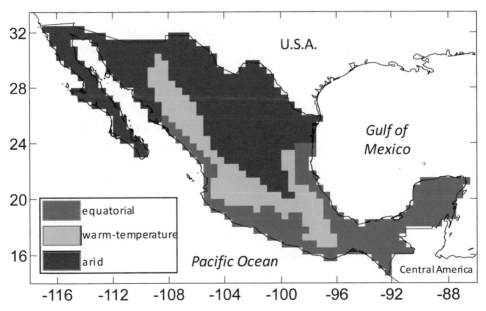

Figure 8.5 Simplified Köppen-Geiger climate classification for Mexico: equatorial, warm temperature, and arid climates.

73

8.3 RESULTS AND DISCUSSION

The droughts calculated with scPDSI for the period 1901-2013 (113 years) are presented in this section. To illustrate the use of the charts designed, the results are shown as follows. First, graphs that are commonly used in this type of analysis are presented. Then, the proposed charts are used. More than a comparison of approaches or graphs, the intention is to show the advantages of polar and radial charts in identifying patterns of drought variation.

As an example of the calculation of the percentage of drought area (PDA), number of clusters (nc), and average distance between clusters (lc), three of the largest droughts are shown in figure 8.6a. These droughts are those of 1953, 1987, and 2011. PDA and lc values are shown rounded for visibility purposes. Drought areas are displayed with dark red colour on each map. These maps are shown from January to December (J to D). For the analysis period, the maximum PDA was 75% occurred in May 2011 (figure 8.6a). This maximum value has a return period of approximately 110 years. Return period was calculated following the methodology of extreme values analysis (EVA), where the maximum annual values are extracted, then the probability of occurrence of the ordered values is calculated. From maps (Figure 8.6a), it is observed that the change of monthly PDA is very smooth over the three years shown in figure 8.6a. In 1953 and 2011, drought areas were located mainly in the northeast and southwest, while in 1987 in the northwest and southeast. The long-term monthly mean values of the analysis period are 27, 27.3, 26.8, 26.4, 25.9, 25.8, 25.8, 26.4, 27, 27.2, 26.9, and 27 %, from January to December, respectively (not shown).

Figure 8.6a shows that March 1953 and December 2011 have approximately PDAs of 55% each, but the March's area consists of 20 clusters (nc), while the second one has four. This example illustrates that nc indicates the degree of aggregation or disaggregation of drought areas. The lc indicates the degree of cluster conglomeration. For example, of the two months mentioned, lc of March 1953 is 11 cells, and for December 2011 is four cells, which implies that areas of December 2011 are closer to each other, as can be seen in figure 8.6a.

Bottom of figure 8.6a displays monthly maps of drought occurrence for 1901-2011 (113 years). Occurrence is calculated as the number of times cells are in drought (details in Chapter 5). The colour scale shows the intervals of occurrence. The highest occurrence is concentrated in the northern region. There are no significant changes between the distribution and the magnitude of the drought occurrence calculated for each month when comparing each month's results.

Figure 8.6b presents the time series of PDA calculated for the whole period. The green boxes indicate the values for the years 1953, 1987 and 2011 shown in figure 8.6a. Figure 8.6b shows values fluctuating over the decades. The decade with the highest number of months of large droughts (PDA ≥ 50%) was 1950. The maximum PDA occurred in 2011. A limitation of this area chart is that identification of months with large values is not straightforward.

The scatter plots of PAD vs nc, PAD vs lc, nc vs lc, and the box-plot of these three characteristics are displayed in figures 8.6c, d, e, and f, respectively. Figure 8.6c shows that when PDA increases the nc also increases until PDA reaches a specific value, then nc begins to decrease. In the case of the lc, it is observed that as the PDA increases the lc decreases. Regarding the nc-lc relationship, it is observed that lc tends to increase as nc does.

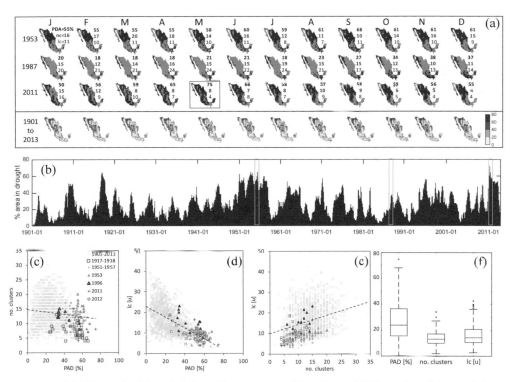

Figure 8.6 (a) Example of the calculation of the percentage of drought area (PDA), number of clusters (nc) and the average distance between clusters (lc) of three selected years. PDA and lc are rounded. (b) PDA time series in an area chart. Green boxes refer to the selected PDAs in 1953, 1987 and 2011. (c) Scatter plots of PAD vs nc, PAD vs lc, nc vs lc, and the box-plot of PDA, nc and lc.

In figures 8.6c, d, and e, years where the maximum PDAs occurred are also displayed. These PDAs occurred in 1953, 1996, 2011 and 2012, which are identified with green triangles, red triangles, blue crosses, and yellow circles, respectively. For these PDAs, different values of nc are exhibited. For example, the PDAs of 1953 (green triangles) are

75

formed with more clusters than those of 2011 (blue crosses), which indicates that the 2011 drought areas are generally larger and contiguous, as shown in figure 8.6a. PDAs of two periods are shown in figures 8.6c, d, and e. These periods are 1917-1918 and 1951-1957, identified with blue squares and pink crosses, respectively. It is observed that in the two selected periods, there are combinations of PDA $\geq$ 50%, $nc \leq 5$ and $lc \leq 5$, which indicate the presence of large drought areas very close to each other. Figures 8.6c and c also show the largest monthly PDA occurred in 2011, which consisted of eight clusters and a very small lc distance that indicates that the large areas were very close to each other. Figure 8.6f shows that the median of PDA, nc and lc time series is approximately 22%, 15, 16 cells, respectively.

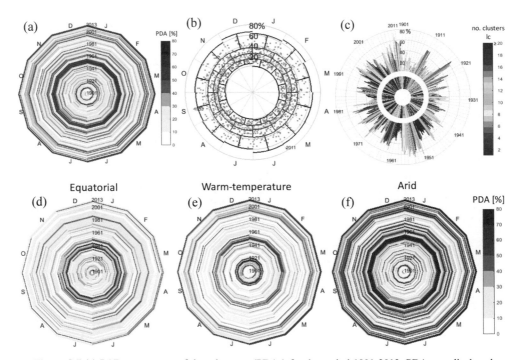

Figure 8.7 (a) PAD, percentage of drought areas (PDAs) for the period 1901-2013. PDAs are displayed from January to December (J to D). (b) Box-plot MOSCA, long-term monthly PDA stats: median (red line), lower and upper edges (blue lines), 25th and 75th percentiles; lower and upper whiskers (black line). Text tags refer to the highest PDAs that occurred in 2011 and 1953. (c) AURA, monthly PDA, nc and lc values from January 1901 to December 2013. All information is displayed clockwise. PAD for the equatorial (d), warm-temperature (e), and arid climate zone (f).

PAD, box-plot MOSCA and AURA are presented in figures 8.7a, b, and c, respectively. PAD shows the PDA time series in the form of rings. The PDAs remain almost constant month by month in most of the years. Years and periods with maximum and minimum

PDA values are observed with dark and white colour, respectively. The 1950s stand out with PDAs $\geq$ 50%. The period of 1917-1918 also shows high PDAs. The 1930s and 1980s have PDAs $\leq$ 40% indicated with whitish rings.

In box-plot MOSCA (figure 8.7b), it is observed that the year-on-year distribution of PDA is practically similar in 12 months. This pattern agrees with the figure 8.6a bottom, where the values of drought occurrence were almost constant throughout the 12 months. The monthly PDA median is approximately 22% in all the cases. Although the distribution of PDA for each month is very similar, it can be seen that the highest values are presented in February, April, May, June and December, reaching 60% (see upper whiskers, figure 8.7b). There are two outliers with PDAs around 75%; the first was in September of 1953 and the second in May of 2011. Drought areas of these two months can be seen in figure 8.6a.

In AURA (figure 8.7c), the distributions of PDA and nc are observed throughout the period. The intervals with PDA $\geq$ 50% and $nc \leq 8$, which indicates the presence of large drought areas close to each other, are the early 1910s, late 1910s, early 1920s, mid-1950s, and the beginning of the previous decade. The distribution of lc is shown in the centre of the AURA diagram. It is observed that in the years where PDA $\geq$ 50% and $nc \leq 8$, the magnitude of lc varies. In the 1950s, lc is $\leq$ 10 cells, which indicates PDAs are constituted by drought areas close to each other in this decade. The same pattern is observed for 1910s. In the first and third quarter of the 20th century, the lowest lc values are observed, while in the second and fourth quarters were the highest lc values. Results for period 2010-2013 show a combination of PDA $\geq$ 40%, $nc \leq 8$ and $lc \leq 10$ cells, which indicates large drought areas close to each other.

Figures 8.7d, e, and f are the PAD charts for the equatorial, warm-temperature, and arid climate zone, respectively (see figure 8.5 for climate zones). It is observed that in each type of climate, the drought areas and their temporal distribution differs. For example, in the equatorial zone's PAD (figure 8.7d), yellow and orange colours are more present that indicate PDAs between 10 and 40%. Semi-rings are also observed, indicating that the monthly PDAs do not remain constant for a given year but change. In the arid zone (figure 8.7f), more rings with dark colours (PDA $\geq$ 60%) are observed. This pattern indicates that the PDAs are higher in this zone than the other two. The rings also show that the monthly PDAs for a given year remain almost constant. The PAD of the warm-temperature zone (figure 8.7e) shows mainly rings and semi-rings from orange to red colour (40% $\leq$ PDA $\leq$ 60%). More white spaces are observed than in the other two zones. This pattern indicates that drought areas that occurred in the equatorial and arid zone were not located at the same time in this warm-temperature zone.

8.4 SUMMARY AND CONCLUSIONS

This chapter introduced a set of visual approaches based on radial and polar charts that aid in interpreting the spatio-temporal variability of drought. The methodology for spatio-temporal drought characterisation was applied on a monthly basis. Percentage of drought area (PDA), number of clusters (nc), and the average distance between clusters (lc) were calculated. The spatio-temporal drought analysis is based on these three features.

The used set of charts reflects the three representations of the spatio-temporal drought variation: (1) Polar Area Diagram (PAD), (2) AnnUal RAdar chart (AURA), and (3) MOnthly Spider ChArt (MOSCA).

Based on the findings, the following conclusions were drawn:

- Radial charts can be used for individual descriptive analysis of drought intensity and trend.

- PAD assists in identifying inter/intra-annual variability of spatio-temporal drought variation and in the detection of drought seasonality.

- AURA helps to detect extreme drought years (outliers) and the number of clusters that contribute to forming the overall drought area. AURA also aids in the analysis of trends.

- MOSCA is suitable to represent the long-term monthly stats of the PDAs. This chart also can assist in the inter-annual comparison of PDAs.

Some points for further research are identified and presented as follows. These polar and radial charts may help to explore how drought moves within the terrestrial hydrological cycle coming from precipitation, i.e. monitor drought propagation (Peters et al. 2003; Van Loon and Van Lanen, 2012). Since the PAD charts resemble tree cross-sections, a possible application of these charts is to analyse sampling three rings data for the assessment of drought impact using ML-based pattern recognition techniques.

9

SPATIAL DROUGHT TRACKING DEVELOPMENT

This chapter is partially based on the following publications.

Diaz, V., Corzo Perez, G. A., Van Lanen, H. A. J., Solomatine, D., and Varouchakis, E. A. (2020). An approach to characterise spatio-temporal drought dynamics. Advances in Water Resources, 137, 103512. https://doi.org/10.1016/j.advwatres.2020.103512

Diaz, V., Corzo Perez, G. A., Van Lanen, H. A. J., Solomatine, D., and Varouchakis, E. A. (2020). Characterisation of the dynamics of past droughts. Science of The Total Environment, 134588. https://doi.org/10.1016/j.scitotenv.2019.134588

9.1 INTRODUCTION

Drought is a regional phenomenon that often covers large territorial extensions (World Meteorological Organisation (WMO), 2006). In terms of the spatial development of the drought, nowadays, available drought monitors deliver information about the spatial extent of droughts (i.e. snapshots) but still, the tracking of these drought areas is lacking, including the temporal variations that form its spatio-temporal dynamics (Hao et al., 2017). This implies that the spatial distribution of drought at a specific time does not give any information about the spatial pathway of the droughts. The implementation of new methodologies to trace drought in space and in time on drought monitors can enhance the spatial tracking and prediction.

The necessity to increase our understanding of the spatio-temporal development of drought has motivated the conduction of studies where drought is considered as a phenomenon that has at least the following main characteristics: duration, intensity (magnitude), and spatial extent (area) (Andreadis et al., 2005; Corzo Perez et al., 2011; Diaz et al., 2018; Herrera-Estrada et al., 2017; Lloyd-Hughes, 2012; Sheffield et al., 2009; Tallaksen et al., 2009; Van Huijgevoort et al., 2013; Vernieuwe et al., 2019). A general framework for the conduction of spatio-temporal analysis of drought can be identified from these studies, which is described as follows. First, a given drought indicator is used to transform the hydro-meteorological variable into water anomalies. The drought indicator is computed in a spatial way, where the study region is embedded in a grid. Then, by establishing a threshold on the drought indicator, the condition of non-drought/drought is identified at each of the cells of the grid. This condition can be expressed in a binary way, i.e. using 0s and 1s. Finally, neighbouring cells showing the same drought condition are aggregated into regions (clusters) by applying a clustering technique. In this way, drought is defined in space and in time, with a spatial extent and duration.

The spatio-temporal analysis of the drought, including the spatial drought tracking, is limited to a few studies such as Diaz et al. (2018), Herrera-Estrada et al. (2017), and Zhou et al. (2019). The first two address the analysis for large-scale studies and the latter presents a basin-scale application. Although there are other publications that consider the study of drought extent locations, they miss the calculation of spatial drought tracks. Following the framework mentioned in the previous paragraph, after the extraction of drought extent (areas), it becomes possible to identify of their location and further construction of the spatial tracks (defined by the linkage between consecutive centroids in time). The calculation and further analysis of these tracks, along with outcome on drought areas, may help to answer the following questions regarding drought dynamics. What are the main places where drought remains? Are there predominant routes followed by drought? How fast does drought change (its extent and location) along its spatial path? Literature review shows that the development of methodologies to describe drought dynamics is still in progress (Chapter 2), therefore more research is needed in this regard (e.g., Herrera-Estrada et al., 2017; Vernieuwe et al., 2019; Zhou et al., 2019).

This chapter introduces the main principles of a new method that complement current drought monitoring by tracking the spatial extent of drought (referred in this document as an area or a cluster). In this study, the description and the application of the methodology to calculate drought tracks are presented in detail. The proposed method is accompanied by an algorithm to calculate the drought characteristics. Both methods are described after this introduction section. The spatio-temporal Contiguous Drought Area (CDA) analysis (Corzo Perez et al., 2011) is used as a basis for the development of the tracking method (Chapter 5). The CDA is applied to identify the neighbouring cells that form the drought clusters. In this chapter, a drought is defined by an onset location, pathway over time, and an end location based on the built tracks. A new drought characteristic is introduced in this study, rotation. Rotation is a feature often calculated when tracking objects in the space dimension. The application of drought tracking method was performed over the country of India for the time period 1901-2013.

9.2 METHODS

9.2.1 S-TRACK: Spatial tracking of drought

The spatial determination of drought tracks was firstly introduced by Diaz et al. (2018) and further developed in this research (Diaz et al., 2020a, 2020b). S-TRACK consists of three main steps: (1) calculation of the spatial drought units (referred to here also as areas or clusters); (2) localisation of centroids; and (3) linkage of centroids (figure 9.1).

Step 1. Spatial drought units computation

In the spatial context, drought units are identified by means of the Contiguous Drought Area (CDA) analysis (Corzo Perez et al., 2011). A CDA is composed of neighbouring cells in drought. As mentioned in the introduction section, the condition of drought or non-drought is indicated with 1s and 0s, respectively, at each cell. Therefore, in each cell drought conditions are indicated when the drought indicator is below or equal to a threshold. Drought indicators (DIs) are mathematical representations of a water anomaly. In general, CDA can be applied over any DI that is in a grid form. Following the CDA methodology, in each time step, the CDAs are computed.

CDA analysis follows a connected-component labelling approach to cluster the cells in drought (Haralick and Shapiro, 1992). In this approach, a two-scan algorithm is applied. Firstly, each cell is numbered for location issues. Then, the first run is performed where the binary grid is explored and connected (contiguous) components (cells) are assigned with provisional labels. These labels point out the connexion of every cell with its 8 nearest neighbours. Within the grid, in a section of 3x3 cells, 9 cells in total, the central cell has 8 surroundings. In this first run, the cell's label does not refer to the number of cluster yet but to the cells with which the given cell is connected. Finally, a second scan is carried out to find similar cell connexions, i.e. clusters. In this second run, clusters are indicated with a unique label. Examination of the grid can be performed indistinctly by columns or by rows. CDA analysis is conducted in each time step over the whole grid. For more details on CDA analysis refer to Corzo Perez et al. (2011).

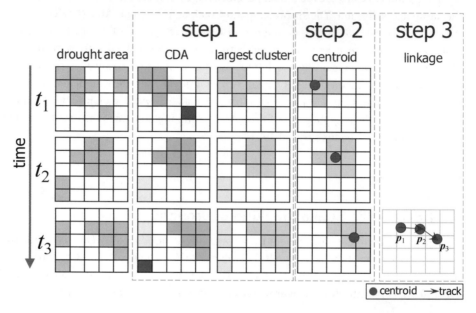

Figure 9.1 Schematic overview of S-TRACK method for spatial drought tracking which involves: (step 1) spatial drought units (clusters) computation, (step 2) centroids localisation, and (step 3) centroids linkage. Example of the procedure is presented for the case of three times steps: from t_1 to t_3. Columns in the diagram show the sequence of the steps. Coloured cells in the first column indicate all cells in drought. Colours in the second column point out different clusters identified. In the third column, the largest contiguous area in drought is presented with a different colour. Only the largest cluster is shown in fourth column and its centroid (p) is indicated by a point. Subscript stands for every time step.

The use of CDA relays on the assumption that the binary description of drought condition (0s and 1s) is homogeneous over the whole grid. Thus, if two or more cells denote drought (value of 1) conditions and are contiguous in space, it is assumed that all of them are part of the same drought unit. In this respect, it is recommended to choose a drought indicator that considers the normalisation of the values in the spatial domain. In this study, a standardised drought indicator was applied as mentioned afterwards, which allowed the clustering of neighbouring cells in drought (cells with 1s).

After the computation of clusters (areas in drought), the major (largest) one is identified in each time step t (figure 9.1). As the tracking algorithm focuses on the calculation of the major spatial drought extent in each time step, small or one-cell units are discriminated with the selection of the largest one, allowing the elimination of possible artefact drought areas.

Step 2. Centroids localisation

After the identification of the major (largest) drought cluster, its centroid (p) is calculated in each time step. This feature is used as the location of the cluster in a similar way as Corzo Perez et al. (2011) and Lloyd-Hughes (2012) presented. The way that clusters are joined in time is explained in the next step. Step 2 and 3 presented in this document, are an extension of the CDA analysis of Corzo Perez et al. (2011). Another point that can be taken into account to indicate the location of a given cluster is, for instance, the one with the lowest drought indicator value (Andreadis et al., 2005; Herrera-Estrada et al., 2017). In this research, we chose the centroid since we already reduce the spatial representation of drought indicator by only 1s and 0s, i.e. drought and non-drought condition, respectively.

Step 3. Centroids linkage

In this step, an algorithm to link centroids of consecutive clusters in time is explained. In this algorithm, a set of rules help to separate or join the sequence in time (figure 9.2). The rules consider two types of threshold parameters: (1) two that control the magnitude (size) of cluster (A, with dimensions L^2), and (2) two that constrain the Euclidean distance between consecutive clusters (Δl, with dimensions L) (figure 9.2). The parameters are denoted as follows: a, b, c and d. The first two used to the drought area A, and the last two to the distance Δl. The output in this step is a time series with 0s and 1s. The time series is denoted by S(t). Here, the value of 1 indicates linkage of clusters in time. If the cluster at time t is not connected with that one at time $t-1$, the value of 0 is used instead. Consecutive values of 1s in the time series S show the occurrence of what is defined as a drought track. The flowchart of the rules for linking the centroids is presented in figure 9.2 and below these rules are explained.

Centroids linkage starts by identifying if the cluster area A is above parameter a (figure 9.2, rule 1). This first comparison helps to discriminate small clusters. If A is below a, there is no connexion between consecutive clusters and this procedure finalised retrieving 0. Before comparing the distance between areas (Δl), the second comparison of A is applied to identify if it is a "very large" area (figure 9.2, rule 2). Parameter b is proposed to consider these large areas alternatively. When A is below b, the parameter c is used to compare distances between clusters (figure 9.2, rule 3). Otherwise, when A is above b, to restrict the distances the parameter d is considered instead (figure 9.2, rule 4). The reason of the second comparison of cluster areas and the use of parameter d is because centroids of clusters with a considerable size may be located farther away from each other and then the distance Δl could fall outside of the limit indicated by parameter c (figure 9.3).

Another parameter that could be included in this linkage algorithm is the degree of overlap between consecutive clusters in time. This way of intersection is not considered directly in our linkage algorithm as a parameter (e.g., percentage of overlapping). The overlap is contemplated in the use of the parameters that control the distance between clusters. An intersection may occur when the distance between centroids is short (figure 9.3).

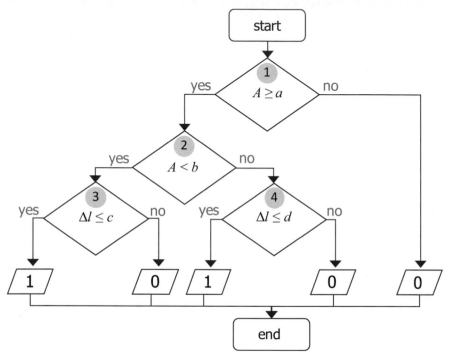

Figure 9.2 Flowchart that shows the rules for linking drought areas (clusters) in time. Numbers in the boxes indicate the sequence of rules 1 to 4. The output of 1 means that the drought area A at time t joins its predecessor at time $t-1$, otherwise output is 0. The distance between the centroids at times t and $t-1$ is represented by Δl. The linking algorithm has the following parameters: a, b, c and d. The first two used to control drought area A, and the last two to the distance Δl.

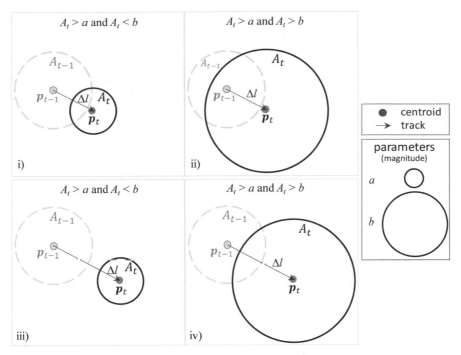

Figure 9.3 Schematic overview of four cases when linking clusters (drought areas) in time. Area at time t is indicated by A_t (bold circle) and its predecessor at time $t-1$ by A_{t-1} (dashed circle). Centroids of areas A_t and A_{t-1} are pointed out by p_t and p_{t-1} (points), respectively. Distance between centroids is represented by Δl (arrow). Centroids in both (i) & (ii) have the same location, in the same way, the centroids in both (iii) & (iv). Areas A_t in (i) & (iii) are of similar size and between the parameters a & b. On the other hand, in (ii) & (iv), areas A_t are also equal but above those parameters (case of a "big" area). Only the parameters of drought area are represented in this figure. Schemes (i) to (iv) help to illustrate the relevance of using parameters that consider not only the magnitude of areas but also the distance between them within the link algorithm. As a distance limit that helps in linking large areas may not be adequate in connecting smaller ones, as shown in (iv) & (iii), two distance parameters are proposed in the link algorithm.

9.2.2 Calculation of drought characteristics

The methodology to build drought tracks allows for identification of paths with an onset and an end location. The information calculated along the paths can help to describe the occurrence of drought. Particularly, it is possible to extract information regarding the duration, severity, as well as rotation. In the following analysis of the spatio-temporal drought dynamics severity has a different meaning compared to on-site analysis or CDA studies: it expresses a certain degree of water missing, an anomaly compared to normal conditions. Herein, severity has a spatial meaning, it is connected to the temporal evolution of the drought are size, irrespective of the strength of the drought. In the following paragraphs, the procedure to calculate drought characteristics is presented. The proposed approach is called DDRASTIC-spatial. DDRASTIC-spatial stands for Drought DuRAtion, SeveriTy and Intensity Computing-spatial events. DDRASTIC-spatial is applied after drought tracks are identified through the S-TRACK algorithm. This approach has as a predecessor (Diaz et al., 2019), a method that lacks the consideration of the elements regarding the spatial domain, such as clusters, locations and paths.

For the calculation of the duration of drought, firstly the onset and an end are obtained. To do so, the time series of 1s and 0s calculated with S-TRACK method is analysed, i.e. $S(t)$. As mentioned, the consecutive sequence of 1s in the time series S, indicate the occurrence of a drought track. One isolated value of 1 shows the linking of two clusters in time. Two consecutive 1 values show the linkage of three clusters in time, and so on. In a sequence of 1s, the time of the first value of 1 (t_{first}) is the time step at which the second and first cluster are connected. The time step of the last value of 1 (t_{last}) is the one when the last and penultimate clusters are linked. The onset ti is defined as $ti = t_{first} - 1$, while the end tf as $tf = t_{last}$. The duration (dd) is calculated with Eq. 9.1.

$$dd = \sum_{t=ti}^{tf} S(t) \qquad \text{(Eq. 9.1)}$$

The magnitudes of areas of the largest clusters calculated in each time step with S-TRACK method are saved in the time series DA (drought area). The drought area is used as the measure of the drought severity. Drought severity (ds) is computed as the sum of drought areas of the period defined by the onset (ti) and the end (tf) (Eq. 9.2). Drought intensity (di) is defined as the ratio between drought severity and duration (Eq. 9.3).

$$ds = \sum_{t=ti}^{tf} DA(t) \qquad \text{(Eq. 9.2)}$$

$$di = ds/dd \qquad \text{(Eq. 9.3)}$$

The definition of locations where a drought path starts and ends can provide its main direction. The initial and final locations are identified using the centroids of the first and last cluster, respectively. The location is a relative position in the spatial domain of the study region. It refers to a point in the axes south-north (S-N) and west-east (W-E) (figure 9.4). The origin of the axes is assigned arbitrarily, here it is proposed to place this origin in the centroid of the study region. The centroid of a particular cluster can be located in one of the nine proposed positions: centre (C), east (E), northeast (NE), north (N), northwest (NW), west (W), southwest (SW), south (S) and southeast (SE) (figure 9.4). Centre (C) is situated in the centroid of the study region (figure 9.4). A point (centroid) is in the centre if the distance (r) between such point and the origin is within the r_{min} radius (figure 9.4). If distance r is out of the r_{min} radius, the location is assigned based on the angle θ. This angle is calculated between the W-E axis and the line defined between the centroid and origin (figure 9.4). All the rules to identify the centroid's location are presented in Table 9.1. Within the algorithm, instead of letters, locations are denoted by means of numerical identifiers (Ids) as presented in the first column of Table 9.1.

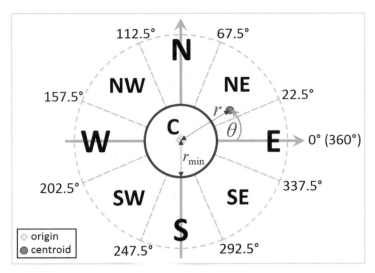

Figure 9.4 Schematic overview of the procedure to define centroid's location of a cluster. A centroid can be located in one of nine positions: centre (C), east (E), northeast (NE), north (N), northwest (NW), west (W), southwest (SW), south (S) and southeast (SE). The symbol r stands for the distance between the cluster's centroid and the one of the region. The angle between W-E axis and the line defined by centroid's cluster is indicated by θ. The radius to define if a cluster is located in the centre (C) of the region is pointed out by r_{min}.

Table 9.1 Rules to define the location of a centroid's cluster. Nine positions are proposed: centre (C), east (E), northeast (NE), north (N), northwest (NW), west (W), southwest (SW), south (S) and southeast (SE).

Id	Rule				Location
0	$r \leq r_{min}$				C
1	$r > r_{min}$ and	$0°$	$\leq \theta <$	$22.5°$ or $337.5°$ $\leq \theta <$ $360°$	E
2	$r > r_{min}$ and	$22.5°$	$\leq \theta <$	$67.5°$	NE
3	$r > r_{min}$ and	$67.5°$	$\leq \theta <$	$112.5°$	N
4	$r > r_{min}$ and	$112.5°$	$\leq \theta <$	$157.5°$	NW
5	$r > r_{min}$ and	$157.5°$	$\leq \theta <$	$202.5°$	W
6	$r > r_{min}$ and	$202.5°$	$\leq \theta <$	$247.5°$	SW
7	$r > r_{min}$ and	$247.5°$	$\leq \theta <$	$292.5°$	S
8	$r > r_{min}$ and	$292.5°$	$\leq \theta <$	$337.5°$	SE

r distance between centroid's cluster and the one of the region.
θ angle between W-E axis and the line defined by centroid's cluster
r_{min} limit distance to consider the location in the centre (C) of the region

Drought tracks provide the whole overview of how drought moves in the spatial domain. Initial and end location (initial and end point of the track) helps to identify the direction followed by a given drought cluster. Another characteristic that complements the description of the dynamics of drought is the rotation. This characteristic is defined as the circular orientation followed by the spatial extent of drought. Rotation is a feature commonly attributed to objects that experience changes in space. It is a basic characteristic analysed in other weather-related phenomena such as cyclones (e.g., Chavas et al., 2017; Rahman et al., 2018) but that has not been investigated much in droughts so far. This study introduces an initial step towards a complete approach for the calculation of drought rotation. This characteristic is included because it is foreseen that it can help analyse drought drivers such as the role played by the climate and land surface control factors on the spatial development of droughts. The drought rotation patterns are expected to be different for each combination of the aforementioned factors.

As the drought track can switch between clockwise and counter-clockwise along the pathway, we propose to classify the rotation in a more general way as (1) mostly clockwise (cw), or (2) mostly counter-clockwise (ccw) (figure 9.5). To determine the rotation, a procedure is considered which makes use of the centroids' coordinates. The algorithm is based on that property to compute a polygon's area (A) from a vector with the coordinates x and y of the vertices (Eq.9. 4). In this area computation algorithm, firstly the sum of products between the coordinates x and y, denoted by ρ (Eq. 9.5), is calculated. Then, ρ is applied to define the rotation direction (Eq. 9.6). In this approach, the coordinates x and y are taken from the ones of centroids' clusters. When there are only two points (two clusters), or when the track is horizontal or vertical, the rotation is not defined, because ρ takes the value of zero. In figure 9.5 two examples of the calculation of rotation are shown for illustration. One example is presented for mostly counter-clockwise (figure 9.5 (left)) and one for mostly clockwise (figure 9.5 (right)). We chose this approach to compute rotation because it distinguishes between "big" and "small" turns in the calculation (Eq. 9.5). The fourth column of tables presented in figure 9.5 provides examples of how the magnitude of each turn is considered differently in the rotation algorithm.

$$A = \frac{1}{2}|\rho| \tag{Eq. 9.4}$$

$$\rho = (x_1 - x_n)(y_1 + y_n) + \sum_{i=1}^{n-1}(x_{i+1} - x_i)(y_{i+1} + y_i) \tag{Eq. 9.5}$$

$$\omega = \begin{cases} \text{cw (mostly clockwise)} & \text{if } \rho > 0 \\ \text{ccw (mostly couter - clockwise)} & \text{if } \rho < 0 \\ \text{nan (not defined)} & \text{if } \rho = 0 \end{cases} \tag{Eq. 9.6}$$

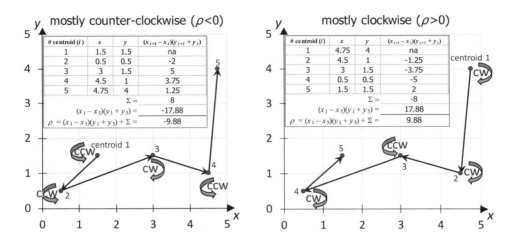

Figure 9.5 Example of rotation calculation. Two types can be assigned: (1) mostly counter-clockwise, when $\rho<0$ (left); and (2) mostly clockwise, when $\rho>0$ (right). The number in each centroid (point) indicates the tracking sequence. Arrows show the track direction, and rotation. Rotation of each line segment is also marked by 'cw' and 'ccw' that stand for clockwise and counter-clockwise, respectively.

9.3 EXPERIMENTAL SETUP

9.3.1 Drought indicator data

Drought tracks were calculated with S-TRACK algorithm for the period 1901 to 2013 (113 years). The analysis was conducted on a monthly basis over India as an example. Data from the Standardized Precipitation Evaporation Index (SPEI) Global Drought Monitor (http://spei.csic.es/) was used (Beguería et al., 2014) to test the proposed methodology for drought tracking and characterisation (Chapter 4). The procedure to calculate SPEI (Vicente-Serrano et al., 2010) is similar to that used to compute the Standardized Precipitation Index (SPI) proposed by Mckee et al. (1993), but taking into account precipitation (P) minus potential evaporation (E) instead of only P. SPEI data from the drought monitor are in a grid form for different temporal aggregation periods. In this study, we used SPEI-6, which corresponds to anomalies of the six-month accumulation of $P - E$. This aggregation usually refers to extended periods of lack of water availability, therefore consequences of a meteorological drought are closer to those caused by a hydrological drought (WMO, 2012).

9.3.2 Drought areas and centroids

Before the application of the drought tracking algorithm, the size of the largest clusters and the distances between the centroids of consecutive clusters in time were calculated. This calculation was performed, on the one hand, to understand the order of their magnitude and frequency, and on the other hand, to set the values of the tracking algorithm parameters.

For the definition of drought areas, usually, the threshold of -1 is used to indicate drought condition in the drought indicators that follow a similar methodology than SPI, also referred as standardised ones. In this research, the same threshold (SPEI=-1) was selected to define drought condition in each cell of the grid in each time step. When SPEI was below -1, with 1s the drought condition was indicated, in another case, with 0s the non-drought status was pointed out. This binary representation allowed the identification of spatial drought units (clusters) through the application of the spatio-temporal analysis of Contiguous Drought Area (CDA) (Sect. 9.2.1).

The largest clusters in each time step were then identified. The area of the largest cluster was compared with the total one to identify the similarity in size between them. It is assumed that the more similar the larger area to the total one, the better the identification of the drought tracks will be. This stands because the tracking algorithm focuses on only one area per time step. For the comparison, the area of all clusters (DA_total) and the area of the largest one (DA_largest) was calculated. Both areas were expressed as percentages calculated as the ratio between the number of cells in drought and the total of them. The total number of cells considered for the mask of India was 1,173.

Once the centroids were identified, the distances between consecutive centroids were calculated over time (Sect 9.2.1). Both the clusters and the distances were calculated for the entire period of analysis on a monthly basis.

9.3.3 Tracking algorithm setup and evaluation

The tracking method can be used in the following two modes: specific and generic.

In specific mode, parameters are calibrated with the use of reported droughts' information. Ideally, this calibration should be carried out considering observed drought tracks. In the absence of this type of data, parameters can be assigned based on the best reproduction of drought occurrence. A drought occurrence refers to the best match of the onset and end time (month) between observed (reported) and calculated droughts.

On the other hand, a way to carry out the generic mode is by following a sensitivity analysis approach. In this mode, the robustness of the structure of the method is examined through the analysis of the outputs under the variation of parameters. Formally sensitivity analysis is performed for the quantification of uncertainty of model results. On the other hand, it can be also applied to evaluate the structure of a model or algorithm (Pannell, 1997). Sensitivity analysis generally allows answering the following questions when evaluating an algorithm. How parameters and output are related? What level of accuracy in the parameters is required? Which parameters are more sensitive? What are the consequences of varying the input parameters?

In this research, both the generic and the specific mode were followed to apply the tracking algorithm. Results are shown in sections 9.3.2 and 9.3.3 for the generic and specific mode, respectively. In the generic mode (Sect. 9.3.2), particularly the effect of parameters was examined over the identification of droughts tracks and characteristics. The questions mentioned in the previous paragraph were used as a guideline to perform such an analysis. For convenience, the four parameters of the S-TRACK algorithm were handled as percentiles.

On the other hand, in the specific mode (Sect. 9.3.3), considering that there is no available information to compare the calculated drought paths in the study area, it was limited to a qualitative analysis of the paths of the most severe droughts reported in the analysis period. The droughts of 1905, 1942, 1965, 1972, 1987, 2000, and 2002 were considered because their severe impacts were referenced (Guha-Sapir, 2019). The qualitative evaluation was focused on the analysis of the extreme incidences using a combination of parameters. From the whole set of combinations we chose three that we consider as key: the one that produces the smallest number of droughts paths (combination_1), the one that yields the largest number (combination_3), as well as the one that produces a number of drought paths similar to the number of years of the analysis period (combination_2). In a more detailed analysis, optimal parameters should be selected based on reported drought' information (source). In the absence of drought tracks, from the source, it is necessary to identify at least the following information: the onset and end month of the reported droughts. The near-optimal parameters are those that provide the best match between the observed and calculated onsets/ends.

9.4 RESULTS

9.4.1 Drought areas and centroids

Drought areas and centroids were computed for the period 1901 to 2013. With respect to the areas, firstly the comparison between the area of all clusters (DA_total) and area of the largest one (DA_largest) was performed. Figure 9.A1 (Annexes) shows the monthly values of both DA_total and DA_largest arranged in matrices. Columns indicate months from January (J) to December (D), while rows point out the year from 1901 to 2013. Drought area value is indicated with a colour scale, where the more red the colour is, the higher the value of the area. The white colour shows months with a small or no value of area. It is observed that DA_total and DA_largest presented similar values. Between both variables, the results showed an agreement in the occurrence of the large values. From the period 1920-1980, DA_largest was slightly smaller than DA_total. The highest values of DA_total and DA_largest were 72 and 70.7%, respectively. Both values occurred in 2001/3. In 1998/3 and 1998/4, no drought areas were identified. The average for the period was 17.4% for the case of DA_total, while for DA_largest was 11.5%. Figure 9.A1

91

(right) presents the difference between DA_total and DA_largest. For all the period, the average of the differences was 5.9%. As DA_largest and DA_total were very similar, it can be considered that the largest cluster is a good proxy to analyse how drought changes in the region without considering the occurrence of two consecutive drought tracks.

The centroids of the largest clusters are presented in figure 9.6. The spatial drought extent is shown schematically with symbols that indicate four intervals of the percentage of drought area with respect to the country extend. The origin of the axes is placed in the centre of the country. It is observed that the spatial distribution of the centroids is almost uniformly distributed over India. However, a higher density of the areas with a considerable extent can be seen in central India.

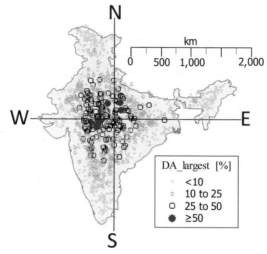

Figure 9.6 Centroids of the largest clusters (DA_largest) identified on a monthly basis. Spatial drought extent is schematised by four symbols pointing out the drought area. The origin of the axes is placed in the centre of the country.

The distances between consecutive clusters in time were calculated also for the whole period. Figure 9.A2 (Annexes) presents the area of the largest cluster (DA_largest) and the distance between clusters (Δl) on a monthly basis for the whole period. It can be observed the occurrence of DA_largest $\geq 25\%$ during all decades of the analysis period. A pattern is observed between DA_largest and Δl: when DA_largest increases, Δl usually tends to decrease. This behaviour was expected, because the more the area increases, the distance between the centroids becomes smaller. This means that the location of the cluster is becoming the same. When Δl does not follow this behaviour, it might be because the consecutive areas in time are very far between them, i.e. they are part of different drought paths.

Figure 9.7 shows the relative frequency of area of the largest cluster (DA_largest) and the distance between clusters (Δl). For both variables, results are displayed in four intervals.

It was observed that as the area increases the frequency of long distances between these areas decreases, while the frequency of small distances increases. For the DA_largest interval of 25-50% and ≥ 50%, the frequency of the small distances (Δl < 250 km) was slightly greater than half of all the distances. This confirms quantitatively what is observed in figure 9.A2: in general when the area grows the distances between the centroids tend to decrease. On the other hand, the small value of the frequency of large distances in large areas (intervals 25-50% and ≥ 50%) indicates that there are large consecutive areas in time that are not necessarily connected to each other.

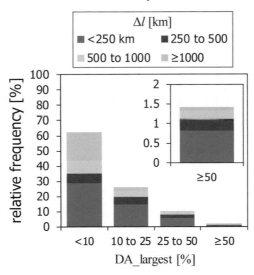

Figure 9.7 Relative frequency of area of the largest cluster (DA_largest) and distances between consecutive clusters in time (Δl).

9.4.2 Drought tracks and characteristics (generic mode)

Based on the distribution of areas and distances between clusters (Sect. 9.3.1), the S-TRACK algorithm was set to take parameters values with the following thresholds: $a ≤ 50$, $b ≥ 50$, $c ≥ 50$, and $d ≥ 50$th percentile (median). As mentioned, a & b are parameters that control the size of clusters (area), and c & d are parameters that restrict the distances between consecutive clusters in time. The average duration, average severity, onset location, end location, were calculated for the different combinations of parameters. Results for a (30, 40, and 50), b (50, 70, and 90), c (50, 60, 70, 80, and 90), and d (50, 60, 70, 80, and 90) are presented in figures 8 and 9.A3 to 9.A7 (Annexes). The a & b parameters are expressed as percentage of drought area and c & d as km.

Figure 9.8 shows the number of drought paths (combination of tracks linked in time). It is observed that the number of drought paths in general increases when a decreases. This is expected since parameter a is the one that allows a cluster to join (or not) consecutive clusters in each time step. When a is small, more clusters are expected to be connected in each time step and therefore more drought paths can be identified. The value of b (used for "very large" areas) seems to influence the number of paths less strictly than a, e.g., when b increases, there is a small proportional increase in the number of paths for all combinations of parameters. The combined variation of b and c seems to influence more the number of paths for small values of d. It is observed that in general, the number of paths drops when a increases and both b, c, and d decrease. In general, the number of drought paths seems to be more sensitive to the changes of parameter a.

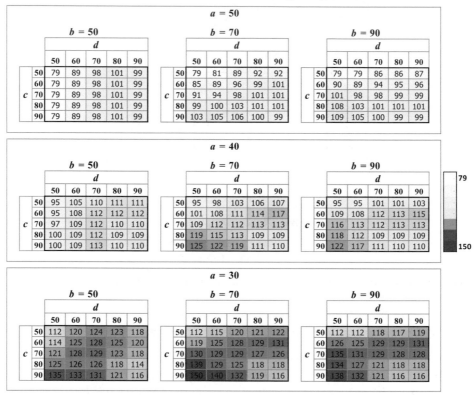

Figure 9.8 Number of drought paths obtained with different combinations of parameters.

In figure 9.A3, average duration of drought paths is presented. Although the variation of average duration seems to be small to the changes of parameters, it is observed a slight increase as a decrease and both b, c and d increase. The average duration seems to be more sensitive to the increase of c and d that are the parameters that control distance between consecutive clusters in time).

Regarding the severity, this value tends to be smaller when a increases and both b, c, and d decrease (figure 9.A4). Severity is calculated as the ratio between the total sum of drought areas and duration (number of months). Its reduction depends on the increase in duration (see Eq. 9.1, 2, and 3). As in the number of drought paths, the average severity is also more sensitive to changes of parameter a. It is observed that when the number of paths decreases, the average severity increases (figures 9.A3 and 9.A4). This is the effect of the selection of a that controls the size of the areas that are joined at each instant of time. If a is small, more areas can be joined and severity may decrease due to the pooling of more areas of small sizes divided by a longer duration (see Eq. 9.1, 2, and 3).

Figures 9.A5 and 9.A6 show the mode of onset and end location of drought paths, respectively. In figure 9.A5, not many changes are observed in the onset location. East seems to be the most common onset location in most combinations of parameters, followed by South. On the other hand, figure 9.A6 shows that the most common end location in most combinations seems to be at the South, followed by East. When both a decreases and b increases, the South tends to be the most common end location.

Figure 9.A7 shows the mode of rotation. It is observed in most cases that mostly clockwise (cw) is the common rotation in the drought paths. When a decreases and b increases, the mostly clockwise rotation seems to be the most common rotation. This is the case when more drought paths were obtained. It is observed that rotation is mostly sensitive to the variations of c and d that are the parameters which control the distance between consecutive clusters in time.

Summary

Table 9.2 shows a summary of how the tracking algorithm responds to different combinations of parameters. In particular, the behaviour of the number of paths, duration, severity, onset and end location, as well as rotation, is indicated. The combinations where it was observed that the values of these characteristics tend to increase or decrease is presented. In general, the most sensitive parameter (important) is the one that controls the minimum area (parameter a). Changes in this parameter have more influence on the result of the number of drought paths and duration. Regarding duration and severity, it is observed that as the paths last longer the severity decreases. This may apply because the severity is calculated as the sum of the areas of clusters that belong to the drought duration. Thus, while the duration increases, the areas that are added tend to be smaller and then the sum does not increase significantly.

The combination 11 (Table 9.2) refers to the identification of paths of "very large" areas. In this combination, it is expected that the initial and final locations will be in the centre. Centroids of these cluster areas tend to be identical to that of the region. For these paths, it is also observed that the rotation tends to be clockwise.

In combinations 6, 7 and 14, by decreasing the parameter that controls the minimum area (parameter a), more drought paths are identified, with the characteristic of being long and with a small severity (formed by a number of smaller areas). In these combinations, drought paths usually start in the East and end in the South, with a clockwise rotation.

If the drought path starts in the South, it usually ends in the East, and in this case, the rotation tends to be counter-clockwise, i.e. the rotation seems to follow the minor turn (combination 14). In other words, if the path starts in the South and ends in the East it is more probable to be directed towards the East showing a counter-clockwise rotation, instead of going first to the West, then North and finally East, showing a clockwise rotation.

Table 9.2 Summary of drought characteristics obtained with different combinations of parameters. Numbers in parentheses indicate the location as presented in figures A5 and A6. Abbreviations ccw and cw stand for counter-clockwise and clockwise, respectively.

#	parameters				number of paths	drought characteristics				
	a	b	c	d		duration	severity	onset location	end location	rotation
1	↑	↓	↓	↓	Decreases, tends to decrease	decreases, tends to decrease	increases			
2	↑	↓	↑	↑	decreases	decreases	increases			
3	↑	↑	↑	↑	decreases	decreases	increases			
4	↑	↑	↓	↓	decreases	decreases	increases			
5	↓	↓	↓	↓	increases	increases	decreases			
6	↓	↓	↑	↑	increases	increases	decreases		tends to south (7)	tends to cw
7	↓	↑	↑	↑	increases	increases, tends to increase	decreases	tends to east (1)		tends to cw
8	↓	↑	↓	↓	increases	increases	decreases			
9	↑	↓	↓	↑	decreases	decreases	increases			
10	↑	↓	↑	↓	decreases	decreases	increases			
11	↑	↑	↓	↑	decreases	decreases	increases, tends to increase	tends to centre (0)	tends to centre (0)	tends to cw
12	↑	↑	↑	↓	decreases	decreases	increases			
13	↓	↓	↓	↑	increases	increases	decreases			
14	↓	↓	↑	↓	increases	increases	decreases	tends to south (7)	tends to east (1)	tends to ccw
15	↓	↑	↓	↑	increases	increases	decreases			
16	↓	↑	↑	↓	increases, tends to increase	increases	decreases, tends to decrease			

9.4.3 Qualitative evaluation (specific mode)

Seven of the most extreme droughts reported during the analysis period were selected to analyse their tracks. These droughts, as it was mentioned earlier, correspond to the following years: 1905, 1942, 1965, 1972, 1987, 2000, and 2002. In the absence of information regarding the dynamics of the droughts, such as trajectories, our validation focuses on the analysis of the calculated tracks in the period when the droughts occurred.

From the whole set of combinations of parameters shown in the previous section, three of them were selected. These were considered key to analyse the results of drought tracks and characteristics. In the first combination, the lowest number of drought paths was obtained in the analysis period (combination_1, a=50, b=50, c=50, d=50). In the second combination there was a number of drought paths similar to the number of years of the analysis period (combination_2, a=40, b=50, c=70, d=80), i.e. more or less a drought path per year. Finally, in the third combination, the highest number of drought paths was identified (combination_3, a=30, b=70, c=90, d=50).

Figure 9.9 presents the occurrence of drought paths calculated for the three combinations of parameters. Columns indicate the months from January (J) to December (D) and the rows show the years. Consecutive cells in colour indicate the occurrence of a drought path (figure 9.9 (top)). The frequency per month was calculated to analyse the distribution of the tracks over the months (figure 9.9 (bottom)). In general, the month with the less frequency of drought tracks was March. From January to July, first part of the year, the frequency is fewer than from August to December. It seems that when the number of drought paths increases they start to occur more frequently in the first part of the year.

Figures of the seven selected droughts were prepared to present the results from the calculation of clusters and distances between centroids to the construction of drought paths. Outputs for the drought of 1987 are presented here in the results section in figure 9.10. For the rest of the droughts, the results are presented in figures 9.A8 to 9.A13 (Annexes). In figure 9.10 (top), clusters and centroids are pointed out from 1987/3 to 1988/6. Areas of largest cluster (DA_largest) and distances between consecutive areas in time (Δl) are shown from 1987/1 to 1988/6 (figure 9.10 (centre)). Duration of the drought paths is indicated in a schematic way with a horizontal line for each combination of parameters. Drought tracks calculated with the three combinations of parameters are also presented (figure 9.10 (bottom)). In most of the seven droughts, the great areas of the largest clusters are in the second half of the year and the first half of the following one (e.g., figure 9.10 (centre)). It was observed that, in general, when DA_largest increases, Δl usually tends to decrease (e.g., figure 9.10 (centre)). This relationship can be explored in further research to define quantitatively the onset and end of the droughts.

Table 9.3 presents a summary of the duration of the selected droughts. It is observed that although the number of drought paths increases from the combination_1 (figure 9.9 (right)) to the combination_3 (figure 9.9 (left)), in terms of the most severe droughts the durations remain almost similar (Table 9.3, column 2 & 4, and figure 9.10 (bottom)). This indicates that more drought tracks were identified in the first part of the year. These drought tracks were not able to join with those of the second half of the previous year. If the parameters c & d that control the distance between centroids are more flexible, i.e. consider longer distances, drought tracks of the second part of the year are more likely to join those of the first part of the next year, as occurs in combination_2. In the combination_2, the drought paths showed the longest durations (Table 9.3, column 3 and figure 9.10 (bottom, centre)).

In all the selected droughts (figures 9.10 and 9.A8 to 9.A13), it is observed that consecutive clusters in time overlap considerably, which suggests that the spatial extent after reaching a considerable size, it remains in the same region. This presence of large drought areas in the same region over time may explain the severity of drought events in those droughts. There is no predominant pathway followed by droughts in those years. In terms of spatial extent, 2000 and 2002 events were the largest as shown in figures 9.A12 and 9.A13, respectively. The drought with the longest duration was that of 1965 (Table 9.3), which is consistent with the reported in (Guha-Sapir, 2019).

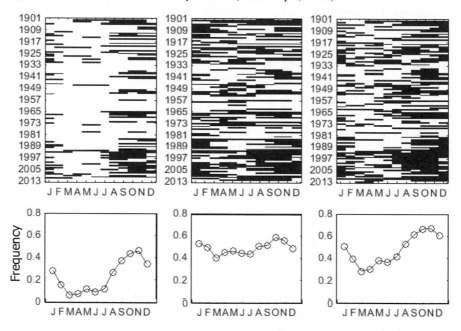

Figure 9.9 Occurrence of drought paths calculated with three combination of parameters: (left) combination_1 (a=50, b=50, c=50, d=50), (centre) combination_2 (a=40, b=50, c=70, d=80), and (right) combination_3 (a=30, b=70, c=90, d=50). Consecutive cells in colour indicate the occurrence of a drought path (top). Frequency is calculated per column from January (J) to December (D) (bottom).

Table 9.3 Duration of selected droughts calculated with three combinations of parameters. In parentheses, the period is indicated.

drought	duration [months]		
	combination_1	combination_2	combination_3
	a=50, b=50, c=50, d=50	a=40, b=50, c=70, d=80	a=30, b=70, c=90, d=50
1	6 (1905/7 to 1905/12)	12 (1905/6 to 1906/5)	6 (1905/7 to 1905/12)
2	5 (1942/10 to 1943/2)	6 (1942/10 to 1943/3)	5 (1942/10 to 1943/2)
3	6 (1965/7 to 1965/12)	22 (1965/5 to 1967/2)	6 (1965/7 to 1965/12)
4	3 (1972/8 to 1972/10)	16 (1972/4 to 1973/7)	3 (1972/8 to 1972/10)
5	6 (1987/9 to 1988/2)	8 (1987/7 to 1988/2)	6 (1987/9 to 1988/2)
6	5 (2000/8 to 2000/12)	11 (2000/8 to 2001/6)	6 (2000/7 to 2000/12)
7	5 (2002/8 to 2002/12)	12 (2002/4 to 2003/3)	6 (2002/8 to 2003/1)

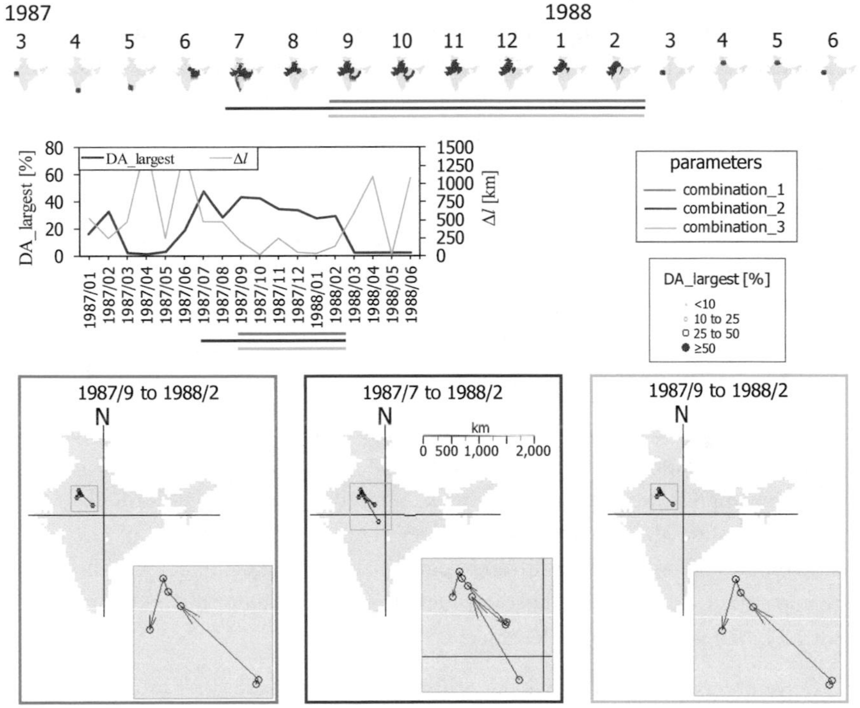

Figure 9.10 Results for drought of 1987. Largest clusters and centroids are indicated from 1987/3 to 1988/6 (top). Area of largest cluster (DA_largest) and distance between consecutive clusters in time (Δl) are displayed from 1987/1 to 1988/6 (centre). The drought duration is pointed out schematically with a horizontal line for each combination of parameters. Drought tracks calculated with the three combinations of parameters are also presented (bottom). Spatial drought extent is schematised by four symbols pointing out the size of area. The origin of the axes is placed in the centre of the country. Arrows point out the direction of each track segment. Insets show zoomed-in views.

9.5 DISCUSSION

9.5.1 Drought indicator and areas

In this initial version of the tracking method, we used a unique threshold over the drought indicator to indicate drought and non-drought conditions in each grid cell (1s and 0s). This threshold is one of the most common proposed in drought studies when considering standardised drought indicators. SPEI drought indicator was applied in this research, but it is possible to use any other indicator, including threshold indicators (Wanders et al., 2010), with the condition of being spatially distributed. The effects of other drought indicator thresholds over the cluster size were not assessed because the scope of this study was limited to testing the drought tracking algorithm.

On the other hand, the clustering algorithm used in this study assumes that all cell values in the space domain are homogeneous. To ensure that this assumption is correct, it is recommended to employ a drought indicator that uses a normalization procedure in its calculation. In addition, our clustering approach is based only on drought indicator values and does not consider others aspects that can influence the spatial extent of drought, such as topography, land use, and climate regions. In further studies, it is recommended to improve the clustering method to incorporate other elements to make it more general. Another way of considering the factors mentioned above, without modifying/changing the clustering algorithm, is the use of a drought indicator that takes into account variables such as soil moisture or runoff.

9.5.2 Drought tracking method

The S-TRACK methodology was performed in both the generic and specific mode explained in the experimental setup section (Sect. 9.2.3.3). On the one hand, in the generic mode, the application was based on the sensitivity analysis of drought tracks and characteristics on the parameters. We presented the performance of the tracks and characteristics when selecting different combinations of parameters. On the other hand, in the specific mode, the analysis of drought tracks of seven of the most severe droughts reported in the literature was presented.

The current version of S-TRACK algorithm focuses on the largest drought areas. In this way, areas with a considerable territorial extent are identified. We are aware that smaller, intense droughts would not be captured by this tracking algorithm. Also, that mild droughts over large areas obtained by the algorithm would overshadow smaller, intense droughts.

Although S-TRACK makes the use of CDA analysis for the extraction of drought clusters, other algorithms used for the same purpose can also be considered. These algorithms include the recursion-based approach (Andreadis et al., 2005; Herrera-Estrada et al., 2017;

Lloyd-Hughes, 2012; Sheffield et al., 2009), and variations of the connected-component labelling approach (Van Huijgevoort et al., 2013; Vernieuwe et al., 2019). The composition of drought clusters extracted with any of these algorithms should be the same. The main difference involves the computational efficiency and processing time among them, which is an important element to consider when processing a large amount of data. In this sense, algorithms based on connected-component labelling are considered more efficient (He et al., 2009).

To connect two consecutive clusters in time and ensure that they are not far in space, the length between centroids of the clusters is taking into account, similar to Herrera-Estrada et al. (2017) and Zhou et al. (2019). The degree of the overlap between these two clusters can be another way to handle the connection between them. A more complete way of joining clusters in time is through the use of the CDA approach but extended to the time domain, i.e. to connect 26 nearest neighbour cells, a cube in the space and time domain, as shown in Corzo Perez et al. (2011), Lloyd-Hughes (2012), and Herrera-Estrada et al. (2017).

In the case that more than one drought track occurs at the same time, i.e. more than one large area present at the same time step, the algorithm will aim to identify the largest one. The algorithm does not detect simultaneous drought tracks, nor merge the areas of the same time step into a single one.

In this research, we compared the area of all clusters and the one of the largest cluster in each time step, to see if the presence of more than one large area is predominant or not. We found that difference between DA_total − DA_largest was, in most of the cases, close to zero (Figure 9.A1). This indicates that the size of the area of the largest cluster is very similar to the total one. Based on the latter, it is assumed that the presence of more than one large cluster at the same time step, is not dominant. Then the research was focused on testing the tracking algorithm, without considering the effect of the presence of more than one simultaneous drought track.

If the presence of more than one consecutive drought track are suspected, an option to perform the tracking method is to test it considering different sub-regions of the study area (analyse it by parts) and then superimpose the drought tracks. In this way, one would expect to identify more than one tracks, if any. In a future version of the tracking algorithm, it is recommended to include the identification of more than one simultaneous drought tracks.

The use of CDA approach can retrieve areas with "islands" of non-drought cells (0s). In this research, we do not consider the possible effects of these holes over the drought tracks construction (the centroid could be located in one of these holes). We assume that centroid is a good spot to locate the contiguous drought area.

101

In largest clusters, the centroid approaches the centroid of the analysis region. This is an expected outcome because if the cluster covers the entire region of analysis the centroid will be similar to that one of the region. In our case, the maximum DA_largest was 70.7%, therefore in the period of analysis, no cluster covered the entire territory. In addition, two simultaneous large clusters are not expected.

Although drought tracks that occurred near the boundaries of the domain may not be calculated well, it is assumed that these tracks do not significantly impact the region because they are at the limits. To improve the calculation in such cases, it is recommended to increase the size of the analysis region.

9.6 SUMMARY AND CONCLUSION

In this study, a method that allows the construction of drought tracks in space was introduced. The onset and end of drought paths (combination of linked drought tracks) are used to compute the drought duration. The information obtained during the path calculation is employed to compute the severity, as well as the onset and end location, direction, and rotation. All these features have been identified as drought characteristics and are framed within the DDRASTIC-spatial methodology, also presented in this document. Outputs of the tracking algorithm S-TRACK and the method for drought characterisation DDRASTIC-spatial help to describe the dynamics of droughts.

S-TRACK has four parameters. Parameters a & b control the size of the cluster (area) to be included in the drought tracks. Parameters c & d limit the distances between consecutive clusters in time (Sect. 9.2.1). In this document, the S-TRACK application was also illustrated in the construction of drought tracks in space.

From the application of S-TRACK, some key findings were presented:

- The number of drought paths, duration, and severity seems to be more sensitive to the change of the parameter that limits the minimum drought area (parameter a) (Sect. 9.2.1).

- If the duration of the drought paths increases, severity does not necessarily do so, because the longer the duration, the areas that make up the path tend to be smaller (Sect. 9.3.2).

- To obtain drought paths with longer durations, it is important to be flexible with the parameters that control the distance between areas (parameters c & i), i.e. to consider larger distances.

102

The outcome of the approach presented in this paper is relevant for (i) drought forecasting, i.e. drought tracks can help to predict how drought moves over a particular region, and (ii) for improving knowledge on drought-generating processes. The first item is more for operational purposes (short term) and the second item – for scientific research (long term).

Regarding the improvement of knowledge on drought-generating processes, i.e. the interaction between climate and land surface characteristics, a new drought characteristic is introduced, the rotation (Sect. 9.2.2). This feature is common in studies of other weather-related phenomena such as cyclones, because it helps in the description/identification of forcing mechanisms behind their spatial development (e.g., Chavas et al., 2017; Rahman et al., 2018). We foresee that this drought characteristic can help in the identification and description of climate and land surface control factors that drive the spatial behaviour of droughts.

For the case of India, we found that consecutive clusters in time overlap considerably in the droughts selected (1905, 1942, 1965, 1972, 1987, 2000, and 2002), which suggests that after reaching a considerable size, the spatial extent of drought remains in the same region. This presence of large drought areas in the same region over time may explain the severity of droughts in those years. There is no predominant pathway followed by droughts in those years. In terms of spatial extent, 2000 and 2002 events were the largest. The drought with the longest duration was that of 1965.

With the presented approach to identify and analyse the drought tracks, the next challenge would be to develop a ML (or hybrid) model able to predict the drought tracks.

10

CONCLUSIONS AND RECOMMENDATIONS

10.1 GENERAL

This study deals with developing and applying analytical and visual methodologies for carrying out the spatio-temporal characterisation of droughts. Droughts are conceptualised as events with a spatial extent, onset and end in space and time, as well as a spatial path composed of the union of successive tracks (Chapter 9). The methodologies developed and presented in this study were built based on computational imaging, ML techniques and approaches combined with the process-based concepts developed in drought studies.

Before the conclusions for each objective are presented, the two hypotheses (H1 and H2) formulated at the beginning of this research are discussed below.

H1) It was hypothesised that the spatio-temporal characterisation of droughts could be improved by including ML techniques in the calculation of droughts. This research showed that incorporating ML techniques, such as clustering, into the identification of drought areas allows for the delineation of the spatial extent, considered in this research as the area of the drought event in a given time step (Chapter 5). The drought areas were further connected over time to build spatial drought trajectories (Chapter 9). Consequently, ML techniques also helped in identifying the largest drought areas, i.e. the most extended ones, which were found to be associated with the most severe drought impacts presented in local reports (Chapter 9).

Recommendations: Regarding the calculation and characterisation of droughts, the following limitations need to be addressed in further developments. First, the algorithm for calculating the connection of drought areas over time, i.e. spatial trajectories, is focused on the largest events (areas), so its application is recommended at the basin or country scale (Chapter 9.5). For continental or global use, an analysis of the sizes of the drought areas must first be conducted to identify whether more than one of the largest drought occurs at the same time. Then, based on this analysis, the continent/globe can be sectioned for individual uses of the algorithm. In future developments, the calculation of more than one drought event should be considered.

H2) It was also hypothesised that drought tracking can be improved by considering the characteristics related to the spatial properties of drought with the appropriate visualisation techniques. Drought tracking was limited to the monitoring of the time series of drought indicators, drought areas or drought impacts. This research presented a method for calculating spatial drought tracks that can enhance drought tracking (Chapter 9).

Recommendations: In further research, including cases of spatial splitting or merging drought areas over time that were not considered in this first realisation of the drought tracking algorithm is recommended. For continental or global applications, the factors that cause drought areas to separate or marge could reveal the details about their development and attenuation. Investigating the drivers with which these spatial behaviours are associated is also necessary. Regarding the visual approaches for analysing drought variation, developing graphs that facilitate the analysis of the spatial trajectories of droughts is necessary. Aspects such as the predominant routes, regions with greater drought activity and the relationship between duration, intensity and trajectory length of the drought are some of the spatial patterns that can be addressed with the help of visual approaches, such as those developed in this thesis (Chapter 8).

The four objectives (O1 to O4, see Section 1.4) set in this research were achieved, and the general conclusions can be summarised as follows:

O1) Methodology for the spatio-temporal characterisation of drought dynamics was developed. This methodology improves the characterisation of drought in space and time by conceptualising drought as an event that develops in space. The spatial features of drought, such as area and location, can be used to describe its dynamics.

O2) Visual approaches were developed to analyse variations in drought characteristics. These approaches were based on radial and polar charts, which facilitated the handling of large amounts of information and the interpretation of drought patterns.

O3) Methodology for monitoring the spatial extent of drought, i.e. drought tracking, was developed and applied to analyse the most extensive droughts at the country level. This methodology can be extended to develop an ML approach for drought prediction.

O4) Applicability of ML techniques to the prediction of crop yield responses to drought was investigated. It was found that drought area is a suitable input for building ML models to predict seasonal crop yield.

In more detail, the conclusions and recommendations are presented below.

10.2 (O1) SPATIO-TEMPORAL CHARACTERISATION OF DROUGHT BASED ON THE PHENOMENON'S SPATIAL FEATURES

Characterising drought, i.e. calculating drought duration, intensity and spatial extent, is often carried out for drought monitoring purposes. Drought is mainly identified (calculated) with the help of drought indicators, as presented in Chapter 2. In a region, e.g. a basin, when the drought indicator is spatially distributed, it is often condensed into a single value for each time step prior to drought calculation. The average is typically used for this aggregation. Although this practice is often widely extended for its simplicity, much of the spatial information is lost in the aggregation (Chapter 6). To avoid this inconvenience of information loss, a methodology based on drought area was tested to compare drought indices. Two comparisons were carried out—one amongst the aggregated values of different indicators and another amongst the calculated drought areas. The methodology was applied at the basin scale, in which meteorological, hydrological and agricultural droughts were analysed (Chapter 6).

The results showed little difference amongst the time series of the area-aggregated drought indicators. However, the drought areas exhibited different behaviours, indicating agreement only between the meteorological drought indicators (SPI and SPEI). The drought area analysis also showed a clearly observable lag in time between the meteorological (SPI and SPEI) and hydrological (SRI) drought indicators. The new drought indicator that incorporated actual evapotranspiration (SEDI), which was developed for this research, showed similar behaviour as SRI, suggesting its use for hydrological drought analysis. Finally, comparisons across the drought areas revealed a seasonal drought variation that was not easily detectable through aggregate drought indicator data.

Recommendations: It is recommended that future studies incorporate analyses of drought propagation, i.e. how water anomalies increase or decrease and how drought duration is modified across precipitation, runoff and soil moisture. Remote sensing variables should also be incorporated to calculate drought indicators in the context of drought area analysis. Forecasting models that use drought areas instead of the aggregated values of drought indicators are recommended.

10.3 (O2) VISUAL APPROACHES TO ANALYSING SPATIO-TEMPORAL DROUGHT VARIATION

Problems can arise when visually assessing the spatio-temporal characteristics of drought. Assessing more than two characteristics at once is not always straightforward because of the difficulty of detecting certain patterns, such as seasonality or periodicity. Another issue is the length of the data period. When information is available for long periods, as in the current case study, which accounts for a century of information (Chapter 8), analysing drought variation is challenging. To deal with these limitations, visual approaches based on radial and polar graphs were developed (see Chapter 8). The results showed that these graphs facilitated the identification of drought variations and the detection of spatial patterns.

Recommendations: Further development of these visual tools is recommended to include other spatial characteristics, such as direction, to help identify the predominant routes of drought. The proposed graphs could aid in exploring the spatial patterns of drought within the framework of big data mining by using a large amount of data from existing drought indicator databases.

10.4 (O3) METHODOLOGY FOR DROUGHT TRACKING

This research introduced a methodology for building the spatial path of a drought, i.e. the union of its successive tracks. The spatial paths of the most extensive droughts in India were calculated for the period between 1901 and 2013 (see Chapter 9). The occurrences of these calculated events were corroborated with documented information from the region. The results indicated that the methodology correctly captured the largest droughts. The data generated from the constructed drought paths, i.e. area and location, were then used to characterise drought dynamics. The droughts were found not to follow a specific spatial path, and those with the greatest spatial extent negatively impacted the country most significantly.

Recommendations: The developed methodology should be used to analyse future droughts across the region. This type of spatio-temporal analysis should be carried out for other large-scale cases. It is also recommended that other drought indicators associated with agricultural drought, such as soil moisture, be calculated.

The results of this research can help improve forecasting in the long run. Further research is recommended to develop a model based on ML techniques that will predict the spatial extent of droughts and track them using the STAND methodology. The ML model's construction may be based on Eq. 10.1 as follows:

$$\left(\Delta L_{t+1}, \theta_{t+1}, da_{t+1} \right) = f\left(\Delta L_t, \theta_t, da_t, \overline{L}, \overline{dd} \right)$$

(Eq. 10.1)

where

da_t = drought area at time t

da_{t+1} = drought area at time $t+1$

ΔL_{t+1} = distance between da_t and da_{t+1}

θ_{t+1} = angle (deg) of the line between the centroids of da_t and da_{t+1}

ΔL_t = distance between da_{t-1} and da_t

θ_t = angle (deg) of the line between the centroids of da_{t-1} and da_t

$\overline{L}$ = average length of trajectories

$\overline{dd}$ = average duration

10.5 (O4) ML MODELS TO PREDICT CROP YIELD RESPONSES TO DROUGHT

Crop yield is one of the variables most used to assess the impact of droughts on agriculture. Crop growth models calculate yield and variables related to plant development, but they are limited in that specific data are needed for computation. Given this limitation, ML models are often widely utilised instead, but their use with the spatial characteristics of droughts as input data is limited.

This research explored the use of drought areas as input data for building a framework to predict seasonal crop yield (Chapter 7). This framework is made up of two components. The first includes polynomial regression (PR) models, and the second considers an ANN. In this framework, the purpose was to compare both types of ML models (PR and ANN) and integrate them into the framework. The logic was as follows: ANN models determine the most accurate predictions, but in practice, issues regarding data retrieval and processing can make the use of equations, i.e. PR, preferable. The proposed framework provides these equations to perform such calculations. The estimates can be further improved when the ANN models are run with new input data. The results indicated that the empirical equations (PR) produced good predictions when using drought area as the input data. ANN provides better estimates, in general.

Recommendations: Future work could include drought areas calculated using remote sensing data. Building models at shorter time scales, such as weeks or even days, is also recommended, as these shorter periods are critical for certain crops.

10.6 CONCLUSION IN BRIEF

This research presented analytical and visual methodologies for carrying out the spatio-temporal characterisation of drought (STAND). The methodologies developed and introduced in this dissertation may be used to help in calculating, monitoring and predicting drought. These methods include the Standardised Evapotranspiration Deficit Index (SEDI) for drought monitoring (Chapter 6); the ML framework for predicting crop yield with drought area as input (Chapter 7); the Polar Area Diagram (PAD), the AnnUal RAdar chart (AURA) and the MOnthly Spider ChArt (MOSCA) for analysing drought variation (Chapter 8); the Drought DuRAtion, SeveriTy and Intensity Computing (DDRASTIC) method for drought characterisation (Chapter 5); and the Spatial TRACKing of drought (S-TRACT) method for drought monitoring (Chapter 9).

The outcome of this research is relevant for the following: 1) drought forecasting (i.e. drought tracks can help predict how spatial drought events move around in a specific region) and 2) improved knowledge of drought-generating processes. The first item is more important for operational purposes (short term), and the second one for scientific research (long term).

The results of this research are expected to help improve the calculation of drought characteristics. A better characterisation of droughts will allow for better development of plans and policies for the management of the negative impacts of droughts. It is hoped that the methodologies for calculating drought and trajectories can also help hydro-meteorological organisations in charge of drought monitoring to deal with this task.

Recommendations for future studies have also been shown. Addressing drought as a spatial event has gained attention in the last decade. In the coming years, studies related to its drivers, the way in which drought spreads through the different components of the hydrological cycle, prediction of drought trajectory and intensity, and spatial patterns of droughts (including spatial splitting and merging), are some of the subjects that are expected to have a fruitful future development in this fascinating topic of drought characterisation and prediction.

REFERENCES

Agutu, N.O., Awange, J.L., Zerihun, A., Ndehedehe, C.E., Kuhn, M., and Fukuda, Y. (2017). Remote Sensing of Environment Assessing multi-satellite remote sensing, reanalysis, and land surface models' products in characterizing agricultural drought in East Africa. *Remote Sensing of Environment*, *194*, 287–302. https://doi.org/10.1016/j.rse.2017.03.041

Andreadis, K.M., Clark, E.A., Wood, A.W., Hamlet, A.F., and Lettenmaier, D.P. (2005). Twentieth-Century Drought in the Conterminous United States. *Journal of Hydrometeorology*, *6*(6), 985–1001. https://doi.org/10.1175/JHM450.1

Bachmair, S, Kohn, I., and Stahl, K. (2015). Exploring the link between drought indicators and impacts. *Nat. Hazards Earth Syst. Sci.*, *15*(6), 1381–1397. https://doi.org/10.5194/nhess-15-1381-2015

Bachmair, Sophie, Stahl, K., Collins, K., Hannaford, J., Acreman, M., Svoboda, M., Knutson, C., Smith, K.H., Wall, N., Fuchs, B., Crossman, N.D., and Overton, I.C. (2016). Drought indicators revisited: the need for a wider consideration of environment and society. *WIREs Water*, *3*(4), 516–536. https://doi.org/10.1002/wat2.1154

Barker, L.J., Hannaford, J., Chiverton, A., and Svensson, C. (2016). From meteorological to hydrological drought using standardised indicators. *Hydrology and Earth System Sciences*, *20*(6), 2483–2505. https://doi.org/10.5194/hess-20-2483-2016

Beguería, S., Vicente-Serrano, S.M., Reig, F., and Latorre, B. (2014). Standardized precipitation evapotranspiration index (SPEI) revisited: parameter fitting, evapotranspiration models, tools, datasets and drought monitoring. *International Journal of Climatology*, *34*(10), 3001–3023. https://doi.org/10.1002/joc.3887

Below, R., Grover-Kopec, E., and Dilley, M. (2007). Documenting Drought-Related Disasters: A Global Reassessment. *J. Environ. Dev.*, *16*(3), 328–344. https://doi.org/10.1177/1070496507306222

Bhalme, H.N. and Mooley, D. a. (1980). Large-Scale Droughts/Floods and Monsoon Circulation. *Monthly Weather Review*, *108*(8), 1197–1211. https://doi.org/10.1175/1520-0493(1980)108<1197:LSDAMC>2.0.CO;2

Bhattacharya, T. and Chiang, J.C.H. (2014). Spatial variability and mechanisms underlying El Niño-induced droughts in Mexico. *Clim. Dyn.*, 3309–3326. https://doi.org/10.1007/s00382-014-2106-8

CENAPRED. (2007). *Drought*. (F. Garcia-Jimenez, O. Fuentes-Mariles, & L. Matias-Ramirez, Eds.). Mexico: CENAPRED.

Chavas, D.R., Reed, K.A., and Knaff, J.A. (2017). Physical understanding of the tropical cyclone wind-pressure relationship. *Nature Communications*, *8*(1), 1360. https://doi.org/10.1038/s41467-017-01546-9

Chlingaryan, A., Sukkarieh, S., and Whelan, B. (2018). Machine learning approaches for crop yield prediction and nitrogen status estimation in precision agriculture: A review. *Computers and Electronics in Agriculture*, *151*(May), 61–69. https://doi.org/10.1016/j.compag.2018.05.012

Corzo Perez, G.A., van Huijgevoort, M.H.J., Voß, F., and van Lanen, H.A.J. (2011). On the spatio-temporal analysis of hydrological droughts from global hydrological models. *Hydrology and Earth System Sciences*, *15*(9), 2963–2978. https://doi.org/10.5194/hess-15-2963-2011

Dai, A. (2011). Characteristics and trends in various forms of the Palmer Drought Severity Index during 1900 – 2008. *Journal of Geophysical Research*, *116*(March), 1–26. https://doi.org/10.1029/2010JD015541

de Brito, M.M. (2021). Compound and cascading drought impacts do not happen by chance: A proposal to quantify their relationships. *Science of The Total Environment*, *778*, 146236. https://doi.org/10.1016/j.scitotenv.2021.146236

Dee, D.P. … Vitart, F. (2011). The ERA-Interim reanalysis: Configuration and performance of the data assimilation system. *Q. J. R. Meteorol. Soc.*, *137*(656), 553–597. https://doi.org/10.1002/qj.828

Diaz Mercado, V., Bâ, K.M., Quentin, E., Ortiz Madrid, F.H., and Gama, L. (2015). Hydrological Model to Simulate Daily Flow in a Basin with the Help of a GIS. *Open Journal of Modern Hydrology*, *05*(03), 58–67. https://doi.org/10.4236/ojmh.2015.53006

Diaz Mercado, V., Corzo Perez, G., Solomatine, D., and van Lanen, H.A.J. (2016). Spatio-temporal Analysis of Hydrological Drought at Catchment Scale Using a Spatially-distributed Hydrological Model. *Procedia Eng.*, *154*, 738–744. https://doi.org/10.1016/j.proeng.2016.07.577

Diaz, V., Corzo, G., Lanen, H.A.J. Van, and Solomatine, D.P. (2019). 4 - Spatiotemporal Drought Analysis at Country Scale Through the Application of the STAND Toolbox. In G. Corzo & E. A. Varouchakis (Eds.), *Spatiotemporal Analysis of Extreme Hydrological Events* (pp. 77–93). Elsevier. https://doi.org/https://doi.org/10.1016/B978-0-12-811689-0.00004-5

Diaz, V., Corzo, G., and Pérez, J.R. (2019). 3 - Large-Scale Exploratory Analysis of the Spatiotemporal Distribution of Climate Projections: Applying the STRIVIng Toolbox. In G. Corzo & E. A. Varouchakis (Eds.), *Spatiotemporal Analysis of Extreme Hydrological Events* (pp. 59–76). Elsevier. https://doi.org/https://doi.org/10.1016/B978-0-12-811689-0.00003-3

Diaz, V., Corzo Perez, G.A., Van Lanen, H.A.J., and Solomatine, D. (2018). *Intelligent drought tracking for its use in Machine Learning: implementation and first results.* (G. La Loggia, G. Freni, V. Puleo, and M. De Marchis, Eds.), *HIC 2018. 13th International Conference on Hydroinformatics* (Vol. 3). Palermo: EasyChair. https://doi.org/10.29007/klgg

Diaz, V., Corzo Perez, G.A., Van Lanen, H.A.J., Solomatine, D., and Varouchakis, E.A. (2020a). An approach to characterise spatio-temporal drought dynamics. *Advances in Water Resources*, *137*, 103512. https://doi.org/https://doi.org/10.1016/j.advwatres.2020.103512

114

Diaz, V., Corzo Perez, G.A., Van Lanen, H.A.J., Solomatine, D., and Varouchakis, E.A. (2020b). Characterisation of the dynamics of past droughts. *Science of The Total Environment*, *718*, 134588. https://doi.org/10.1016/j.scitotenv.2019.134588

Elshorbagy, A., Corzo, G., Srinivasulu, S., and Solomatine, D.P. (2010). Experimental investigation of the predictive capabilities of data driven modeling techniques in hydrology - Part 2: Application. *Hydrology and Earth System Sciences*, *14*(10), 1943–1961. https://doi.org/10.5194/hess-14-1943-2010

Farahmand, A. and AghaKouchak, A. (2015). A generalized framework for deriving nonparametric standardized drought indicators. *Adv. Water Resour.*, *76*, 140–145. https://doi.org/10.1016/j.advwatres.2014.11.012

Florescano, E., Sancho Y Cervera, J., and Perez, D. (1980). Las sequías en México : historia , características y efectos. *Comer. Exter.*, *30*(7), 747–757.

Food and Agriculture Organization of the United Nations (FAO). (2017). *The Impact of disasters and crises on agriculture and Food Security*. Retrieved from www.fao.org/publications

Food and Agriculture Organization of the United Nations (FAO) and Robert B Daugherty Water for Food Institute at the University of Nebraska. (2015). *Yield gap analysis of field crops, Methods and case studies*. (V. O. Sadras, K. G. G. Cassman, P. Grassini, A. J. Hall, W. G. M. Bastiaanssen, A. G. Laborte, ... P. Steduto, Eds.), *FAO Water Reports* (Vol. 41). Rome, Italy.

Ghosh, K., Balasubramanian, R., Bandopadhyay, S., Chattopadhyay, N., Singh, K.K., and Rathore, L.S. (2014). Development of crop yield forecast models under FASAL-a case study of kharif rice in West Bengal. *Journal of Agrometeorology*, *16*(1), 1–8.

Govindaraju, R.S. (2000). Artificial Neural Networks in Hydrology. I: Preliminary Concepts. *Journal of Hydrologic Engineering*, *5*(2), 115–123. https://doi.org/10.1061/(ASCE)1084-0699(2000)5:2(115)

Govindaraju, R.S. (2013). Special Issue on Data-Driven Approaches to Droughts. *J. Hydrol. Eng.*, *18*(7), 735–736. https://doi.org/10.1061/(ASCE)HE.1943-5584.0000812

Guha-Sapir, D. (2019). EM-DAT: The Emergency Events Database - Université catholique de Louvain (UCL) - CRED. Retrieved from www.emdat.be

Hannaford, J., Lloyd-Hughes, B., Keef, C., Parry, S., and Prudhomme, C. (2011). Examining the large-scale spatial coherence of European drought using regional indicators of precipitation and streamflow deficit. *Hydrological Processes*, *25*(7), 1146–1162. https://doi.org/10.1002/hyp.7725

Hao, Z and Singh, V. (2012). Entropy-based method for bivariate drought analysis. *Journal of Hydrologic Engineering*, *18*(7), 780–786. https://doi.org/10.1061/(ASCE)HE.1943-5584.0000621.

Hao, Zengchao, Yuan, X., Xia, Y., Hao, F., and Singh, V.P. (2017). An Overview of Drought Monitoring and Prediction Systems at Regional and Global Scales. *Bulletin of the American Meteorological Society*, *98*(9), 1879–1896. https://doi.org/10.1175/BAMS-D-15-00149.1

Haralick, R.M. and Shapiro, L.G. (1992). *Computer and Robot Vision, Volume I.* Addison-Wesley.

Harding, R., Best, M., Blyth, E., Hagemann, S., Kabat, P., Tallaksen, L.M., Warnaars, T., Wiberg, D., Weedon, G.P., Lanen, H. Van, Ludwig, F., and Haddeland, I. (2011). WATCH: Current Knowledge of the Terrestrial Global Water Cycle. *J. Hydrometeorol.*, *12*(6), 1149–1156. https://doi.org/10.1175/JHM-D-11-024.1

Harris, I., Jones, P.D., Osborn, T.J., and Lister, D.H. (2014). Updated high-resolution grids of monthly climatic observations - the CRU TS3.10 Dataset. *Int. J. Climatol.*, *34*(3), 623–642. https://doi.org/10.1002/joc.3711

He, L., Chao, Y., Suzuki, K., and Wu, K. (2009). Fast Connected-component Labeling. *Pattern Recogn.*, *42*(9), 1977–1987. https://doi.org/10.1016/j.patcog.2008.10.013

Herrera-Estrada, J.E., Satoh, Y., and Sheffield, J. (2017). Spatio-Temporal Dynamics of Global Drought. *Geophysical Research Letters*, 2254–2263. https://doi.org/10.1002/2016GL071768

Huffman, G.J., Adler, R.F., Bolvin, D.T., and Gu, G. (2009). Improving the global precipitation record: GPCP Version 2.1. *Geophysical Research Letters*, *36*(17), 1–5. https://doi.org/10.1029/2009GL040000

Jeb Bell, Gleen Tootle, Larry Pochop, Greg Kerr, R.S. (2012). Drought Analysis under Climate Change Using Copula. *Journal of Hydrologic Engineering*, *18*(7), 746–759. https://doi.org/10.1061/(ASCE)HE.1943-5584

Kao, S.C. and Govindaraju, R.S. (2010). A copula-based joint deficit index for droughts. *J. Hydrol.*, *380*(1–2), 121–134. https://doi.org/10.1016/j.jhydrol.2009.10.029

Keyantash, J. and Dracup, J.A. (2002). The quantification of drought: An evaluation of drought indices. *Bulletin of the American Meteorological Society.* https://doi.org/10.-0477

Kim, T. woong and Valdes, J. (2003). A nonlinear model for drought forecasting based on conjunction of wavelet rransformations and neural networks. *J. Hydrol. Eng.*, *0072*(6), 319–328. https://doi.org/10.1061/(ASCE)1084-0699(2003)8:6(319)

Kim, W., Iizumi, T., and Nishimori, M. (2019). Global Patterns of Crop Production Losses Associated with Droughts from 1983 to 2009. *Journal of Applied Meteorology and Climatology*, *58*(6), 1233–1244. https://doi.org/10.1175/JAMC-D-18-0174.1

Kogan, F.N. (1995). Application of vegetation index and brightness temperature for drought detection. *Adv. Sp. Res.*, *15*(11), 91–100. https://doi.org/10.1016/0273-1177(95)00079-T

Kottek, M., Grieser, J., Beck, C., Rudolf, B., and Rubel, F. (2006). World map of the Köppen-Geiger climate classification updated. *Meteorol. Zeitschrift*, *15*(3), 259–263. https://doi.org/10.1127/0941-2948/2006/0130

Krzywinski, M., Schein, J., Birol, I., Connors, J., Gascoyne, R., Horsman, D., Jones, S.J., and Marra, M.A. (2009). Circos: An information aesthetic for comparative genomics. *Genome Research*, *19*(9), 1639–1645. https://doi.org/10.1101/gr.092759.109

Lloyd-Hughes, B. (2012). A spatio-temporal structure-based approach to drought characterisation. *International Journal of Climatology*, *32*(3), 406–418. https://doi.org/10.1002/joc.2280

Maier, H.R. and Dandy, G.C. (2000). Neural networks for the prediction and forecasting of water resources variables: a review of modelling issues and applications. *Environmental Modelling & Software*, *15*(1), 101–124. https://doi.org/10.1016/S1364-8152(99)00007-9

Mallya, G. and Tripathi, S. (2013). Probabilistic assessment of drought characteristics using a hidden Markov model. *J. Hydrol. ...*, *18*(7), 834–845. https://doi.org/10.1061/(ASCE)HE.1943-5584.0000699.

Markonis, Y., Efstratiadis, A., Koukouvinos, A., Mamassis, N., and Koutsoyiannis, D. (2013). *Investigation of drought characteristics in different temporal and spatial scales: a case in the Mediterranean region.*

Maskey, S. and Trambauer, P. (2014). Hydrological Modeling for Drought Assessment. *Hydro-Meteorological Hazards, Risks, and Disasters*, 263–282. https://doi.org/10.1016/B978-0-12-394846-5.00010-2.

May, R., Dandy, G., and Maier, H. (2011). Review of Input Variable Selection Methods for Artificial Neural Networks. In G. Dandy (Ed.), *Artificial Neural Networks - Methodological Advances and Biomedical Applications* (p. Ch. 2). Rijeka: InTech. https://doi.org/10.5772/16004

Mckee, T.B., Doesken, N.J., and Kleist, J. (1993). The relationship of drought frequency and duration to time scales. *AMS 8th Conf. Appl. Climatol.*, (January), 179–184. https://doi.org/citeulike-article-id:10490403

Mishra, A., Desai, V., and Singh, V. (2007). Drought Forecasting Using a Hybrid Stochastic and Neural Network Model. *J. Hydrol. Eng.*, *12*(6), 626–638. https://doi.org/10.1061/(ASCE)1084-0699(2007)12:6(626)

Mishra, A.K. and Singh, V.P. (2010). A review of drought concepts. *Journal of Hydrology*, *391*(1–2), 202–216. https://doi.org/10.1016/j.jhydrol.2010.07.012

Mishra, A.K. and Singh, V.P. (2011). Drought modeling - A review. *J. Hydrol.*, *403*(1–2), 157–175. https://doi.org/10.1016/j.jhydrol.2011.03.049

Modanesi, S., Massari, C., Camici, S., Brocca, L., and Amarnath, G. (2020). Do Satellite Surface Soil Moisture Observations Better Retain Information About Crop-Yield Variability in Drought Conditions? *Water Resources Research*, *56*(2), 0–3. https://doi.org/10.1029/2019WR025855

Montesino Pouzols, F. and Lendasse, A. (2010). Effect of different detrending approaches on computational intelligence models of time series. In *The 2010 International Joint Conference on Neural Networks (IJCNN)* (pp. 1–8). IEEE. https://doi.org/10.1109/IJCNN.2010.5596314

Narasimhan, B. and Srinivasan, R. (2005). Development and evaluation of Soil Moisture Deficit Index (SMDI) and Evapotranspiration Deficit Index (ETDI) for agricultural drought monitoring. *Agric. For. Meteorol.*, *133*(1–4), 69–88. https://doi.org/10.1016/j.agrformet.2005.07.012

Naresh Kumar, M., Murthy, C.S., Sesha Sai, M.V.R., and Roy, P.S. (2012). Spatiotemporal analysis of meteorological drought variability in the Indian region using standardized precipitation index. *Meteorological Applications*, *19*(2), 256–264. https://doi.org/10.1002/met.277

Naumann, G., Dutra, E., Barbosa, P., Pappenberger, F., Wetterhall, F., and Vogt, J. V. (2014). Comparison of drought indicators derived from multiple data sets over Africa. *Hydrology and Earth System Sciences*, *18*(5), 1625–1640. https://doi.org/10.5194/hess-18-1625-2014

Palmer, W.C. (1965). Meteorological Drought. *U.S. Weather Bur. Res. Pap. No. 45.* Retrieved from https://www.ncdc.noaa.gov/temp-and-precip/drought/docs/palmer.pdf

Pannell, D.J. (1997). Sensitivity analysis of normative economic models: theoretical framework and practical strategies. *Agricultural Economics*, *16*(2), 139–152. https://doi.org/https://doi.org/10.1016/S0169-5150(96)01217-0

Peters, E., Bier, G., van Lanen, H.A.J., and Torfs, P.J.J.F. (2006). Propagation and spatial distribution of drought in a groundwater catchment. *Journal of Hydrology*, *321*(1–4), 257–275. https://doi.org/10.1016/j.jhydrol.2005.08.004

Peters, E., Torfs, P.J.J.F., van Lanen, H.A.J., and Bier, G. (2003). Propagation of drought through groundwater—a new approach using linear reservoir theory. *Hydrol. Process.*, *17*(15), 3023–3040. https://doi.org/10.1002/hyp.1274

Prudhomme, C. and Sauquet, E. (2007). *Modelling a regional drought index in France.* Retrieved from http://nora.nerc.ac.uk/1366/

Rahman, S.M., Yang, R., and Di, L. (2018). Clustering Indian Ocean Tropical Cyclone Tracks by the Standard Deviational Ellipse. *Climate.* https://doi.org/10.3390/cli6020039

Rahmati, O., Falah, F., Dayal, K.S., Deo, R.C., Mohammadi, F., Biggs, T., Moghaddam, D.D., Naghibi, S.A., and Bui, D.T. (2020). Machine learning approaches for spatial modeling of agricultural droughts in the south-east region of Queensland Australia. *Science of the Total Environment*, *699*, 134230. https://doi.org/10.1016/j.scitotenv.2019.134230

Reynolds, C.A., Yitayew, M., Slack, D.C., Hutchinson, C.F., Huete, A., and Petersen, M.S. (2000). Estimating crop yields and production by integrating the FAO Crop Specific Water Balance model with real-time satellite data and ground-based ancillary data. *International Journal of Remote Sensing*, *21*(18), 3487–3508. https://doi.org/10.1080/014311600750037516

Sawasawa, H. (2003). *Crop yield estimation: Integrating RS, GIS and management factors, a case study of Birkoor and Kortigiri Mandals. MSc thesis.* International Institute for Geo-information Science and Earth Observation. Retrieved from http://www.itc.nl/library/papers_2003/msc/nrm/sawasawa.pdf

Sheffield, J. and Wood, E.F. (2011). *Drought: Past problems and future scenarios.* (P. Earthscan, Ed.). London.

Sheffield, Justin, Andreadis, K.M., Wood, E.F., and Lettenmaier, D.P. (2009). Global and continental drought in the second half of the twentieth century: Severity-area-duration analysis and temporal variability of large-scale events. *J. Clim.*, *22*(8), 1962–1981. https://doi.org/10.1175/2008JCLI2722.1

Shin, H.S. and Salas, J.D. (2000). Regional drought analysis based on neural networks. *Http://Dx.Doi.Org/10.1061/(ASCE)1084-0699(2000)5:2(145)*, *5*(2), 145–155.

Shukla, S. and Wood, A.W. (2008). Use of a standardized runoff index for characterizing hydrologic drought. *Geophys. Res. Lett.*, *35*(2), 1–7. https://doi.org/10.1029/2007GL032487

Smith, K.A., Barker, L.J., Tanguy, M., Parry, S., Harrigan, S., and Legg, T.P. (2019). A multi-objective ensemble approach to hydrological modelling in the UK: an application to historic drought reconstruction, 3247–3268.

Solomatine, D.P. and Ostfeld, A. (2008). Data-driven modelling: some past experiences and new approaches. *Journal of Hydroinformatics*, *10*(1), 3. https://doi.org/10.2166/hydro.2008.015

Stagge, J.H., Tallaksen, L.M., Gudmundsson, L., Van Loon, A.F., and Stahl, K. (2015). Candidate Distributions for Climatological Drought Indices (SPI and SPEI). *Int. J. Climatol.*, *35*(13), 4027–4040. https://doi.org/10.1002/joc.4267

Talabis, M.R.M., McPherson, R., Miyamoto, I., Martin, J.L., and Kaye, D. (2015). Chapter 1 - Analytics defined. In M. R. M. Talabis, R. McPherson, I. Miyamoto, J. L. Martin, & D. B. T. I. S. A. Kaye (Eds.), *Information Security Analytics* (pp. 1–12). Boston: Syngress. https://doi.org/https://doi.org/10.1016/B978-0-12-800207-0.00001-0

Tallaksen, L. M. and Van Lanen, H.A.J. (2004). *Hydrological Drought - Processes and Estimation Methods for Streamflow and Groundwater. Developments in Water Sciences 48*. (L. M. Tallaksen & H. A. J. Van Lanen, Eds.). The Netherlands: Elsevier B.V.

Tallaksen, Lena M, Hisdal, H., and Lanen, H.A.J. Van. (2009). Space–time modelling of catchment scale drought characteristics. *Journal of Hydrology*, *375*(3–4), 363–372. https://doi.org/10.1016/j.jhydrol.2009.06.032

Tase, N. (1976). *Area-deficit-intensity characteristics of droughts*. Colo. State Univ., Fort Collins, Colo., Hydrol. Pap. 87.

Trambauer, P., Maskey, S., Werner, M., Pappenberger, F., Van Beek, L.P.H., and Uhlenbrook, S. (2014). Identification and simulation of space-time variability of past hydrological drought events in the Limpopo River basin, southern Africa. *Hydrol. Earth Syst. Sci.*, *18*(8), 2925–2942. https://doi.org/10.5194/hess-18-2925-2014

Udmale, P., Ichikawa, Y., Ning, S., Shrestha, S., and Pal, I. (2020). A statistical approach towards defining national-scale meteorological droughts in India using crop data. *Environmental Research Letters*, *15*(9). https://doi.org/10.1088/1748-9326/abacfa

United Nations Office for Disaster Risk Reduction. (2021). *GAR Special Report on Drought 2021*. Geneva, Switzerland.

Van Der Schrier, G., Barichivich, J., Briffa, K.R., and Jones, P.D. (2013). A scPDSI-based global data set of dry and wet spells for 1901-2009. *Journal of Geophysical Research Atmospheres*, *118*(10), 4025–4048. https://doi.org/10.1002/jgrd.50355

van Huijgevoort, M.H.J., Hazenberg, P., van Lanen, H. a. J., Teuling, a. J., Clark, D.B., Folwell, S., Gosling, S.N., Hanasaki, N., Heinke, J., Koirala, S., Stacke, T., Voss, F., Sheffield, J., and Uijlenhoet, R. (2013). Global Multimodel Analysis of Drought in Runoff for the Second Half of the Twentieth Century. *J. Hydrometeorol.*, *14*(5), 1535–1552. https://doi.org/10.1175/JHM-D-12-0186.1

van Klompenburg, T., Kassahun, A., and Catal, C. (2020). Crop yield prediction using machine learning: A systematic literature review. *Computers and Electronics in Agriculture*, *177*(July), 105709. https://doi.org/10.1016/j.compag.2020.105709

Van Loon, A. F. and Van Lanen, H.A.J. (2012). A process-based typology of hydrological drought. *Hydrology and Earth System Sciences*, *16*(7), 1915–1946. https://doi.org/10.5194/hess-16-1915-2012

Van Loon, Anne F. (2015). Hydrological drought explained. *Wiley Interdisciplinary Reviews: Water*, *2*(4), 359–392. https://doi.org/10.1002/wat2.1085

Velasco, I. (2012). *Sequía y Cambio Climático* (Vol. 1). ISBN 9786077563501. Mexico

Vernieuwe, H., De Baets, B., and Verhoest, N.E.C. (2020). A mathematical morphology approach for a qualitative exploration of drought events in space and time. *International Journal of Climatology*, *40*(1), 530–543. https://doi.org/10.1002/joc.6226

Vicente-Serrano, S.M., Beguería, S., and López-Moreno, J.I. (2010). A Multiscalar Drought Index Sensitive to Global Warming: The Standardized Precipitation Evapotranspiration Index. *Journal of Climate*, *23*(7), 1696–1718. https://doi.org/10.1175/2009JCLI2909.1

Vicente-Serrano, S.M., Beguería, S., Lorenzo-Lacruz, J., Camarero, J.J., López-Moreno, J.I., Azorin-Molina, C., Revuelto, J., Morán-Tejeda, E., and Sanchez-Lorenzo, A. (2012). Performance of drought indices for ecological, agricultural, and hydrological applications. *Earth Interact.*, *16*(10). https://doi.org/10.1175/2012EI000434.1

Wanders, N., van Lanen, H.A.J., and van Loon, A.F. (2010). *Indicators for Drought Characterization on a Global Scale. Watch Tech. Rep. No . 24*. Retrieved from http://www.eu-watch.org/media/default.aspx/emma/org/10646416/WATCH+Technical+Report+Number+24+Indicators+For+Drought+Characterization+on+a+Global+Scale.pdf

Weedon, G.P., Gomes, S., Viterbo, P., Shuttleworth, W.J., Blyth, E., Österle, H., Adam, J.C., Bellouin, N., Boucher, O., and Best, M. (2011). Creation of the WATCH Forcing Data and Its Use to Assess Global and Regional Reference Crop Evaporation over Land during the Twentieth Century. *J. Hydrometeorol.*, *12*(5), 823–848. https://doi.org/10.1175/2011JHM1369.1

Wells, N., Goddard, S., and Hayes, M.J. (2004). A self-calibrating Palmer Drought Severity Index. *Journal of Climate*, *17*(12), 2335–2351. https://doi.org/10.1175/1520-0442(2004)017<2335:ASPDSI>2.0.CO;2

White, M.A., Thornton, P.E., and Running, S.W. (1997). A continental phenology model for monitoring vegetation responses to interannual climatic variability. *Global Biogeochemical Cycles*, *11*(2), 217–234. https://doi.org/10.1029/97GB00330

Wilhite, D.A. (Ed). (2000). Drought as a natural hazard: concepts and definitions. In *DROUGHT, A Global Assessment, Vol I and II, Routledge Hazards and Disasters Series*. Routledge, London.

Wong, G., Lambert, M.F., Leonard, M., and Metcalfe, a V. (2010). Drought Analysis Using Trivariate Copulas Conditional on Climatic States. *Journal of Hydrologic Engineering*, *15*(2), 129–141. https://doi.org/Doi 10.1061/(Asce)He.1943-5584.0000169

Wong, G., van Lanen, H. a. J., and Torfs, P.J.J.F. (2013). Probabilistic analysis of hydrological drought characteristics using meteorological drought. *Hydrol. Sci. J.*, *58*(2), 253–270. https://doi.org/10.1080/02626667.2012.753147

World Meteorological Organization (WMO). (2006). *Drought monitoring and early warning: concepts, progress and future challenges. WMO-No. 1006*. Geneva, Switzerland. Retrieved from http://www.droughtmanagement.info/literature/WMO_drought_monitoring_early_warning_2006.pdf

World Meteorological Organization (WMO). (2011). *Agricultural Drought Indices. Proceedings of the WMO/UNISDR Expert Group Meeting on Agricultural Drought Indices, 2-4 June 2010, Murcia, Spain*. (M. V. K. Sivakumar, R. P. Motha, D. A. Wilhite, & D. A. Wood, Eds.) (AGM-11, WM). Retrieved from http://www.whycos.org/WMO/clw/agm/documents/AgriculturalDroughtIndicesProceedings29311.pdf

World Meteorological Organization (WMO). (2012). *Standardized Precipitation Index user guide. WMO-No. 1090*. Geneva, Switzerland. Retrieved from http://library.wmo.int/pmb_ged/wmo_1090_en.pdf

World Meteorological Organization (WMO) and Global Water Partnership (GWP). (2016). Handbook of drought indicators and indices (M. Svoboda and B.A. Fuchs). Integrated drought management programme (IDMP), Integrated drought management tools and guidelines Series 2. *WMO-No. 1173*. Geneva, Switzerland.

Wu, X., Vuichard, N., Ciais, P., Viovy, N., Wang, X., Magliulo, V., and Wattenbach, M. (2016). ORCHIDEE-CROP (v0), a new process-based agro-land surface model: model description and evaluation over Europe, 857–873. https://doi.org/10.5194/gmd-9-857-2016

Yevjevich, V. (1967). *An objective approach to definitions and investigations of continental hydrologic droughts. Hydrology Paper 23* (Vol. 23). Fort Collins, Colorado. Retrieved from https://dspace.library.colostate.edu/bitstream/handle/10217/61303/HydrologyPapers_n23.pdf

Yevjevich, V. and Karplus, A.K. (1973). *Area-time structure of the monthly precipitation process*. Colo. State Univ., Fort Collins, Colo., Hydrol. Pap. 64.

Zaidman, M.D., Rees, H.G., and Young, A.R. (2001). Spatio-temporal development of streamflow droughts in north-west Europe, *5*(4), 733–751.

Zaidman, M.D., Rees, H.G., and Young, A.R. (2002). Spatio-temporal development of streamflow droughts in north-west Europe. *Hydrology and Earth System Sciences*, *6*(4), 733–751. https://doi.org/10.5194/hess-6-733-2002

Zhou, H., Liu, Y., and Liu, Y. (2019). An Approach to Tracking Meteorological Drought Migration. *Water Resources Research*, *55*(4), 3266–3284. https://doi.org/10.1029/2018WR023311

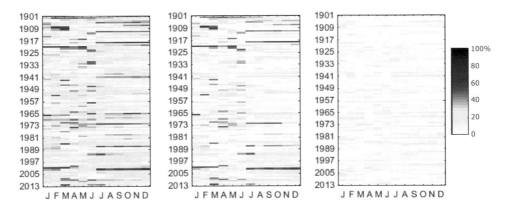

Figure A1 Monthly values of drought area considering all clusters (DA_total, left), and considering only the largest one (DA_largest, centre). Right panel shows the difference between DA_total and DA_largest.

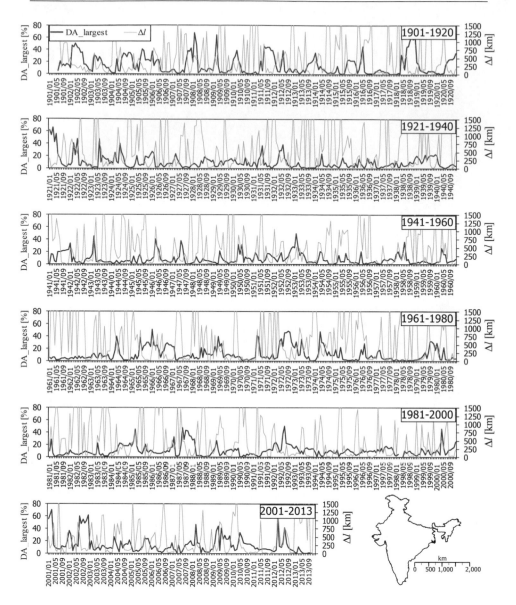

Figure A2 Area of the largest cluster (DA_largest) and distances between consecutive centroids in time (Δ*l*) for the period 1901-2013.

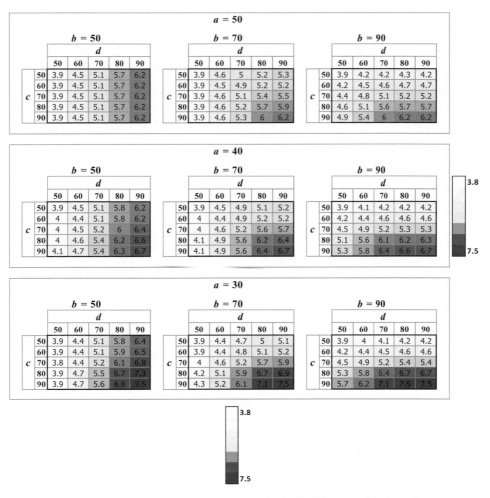

Figure A3 Average duration (months) of drought paths obtained with different combinations of parameters.

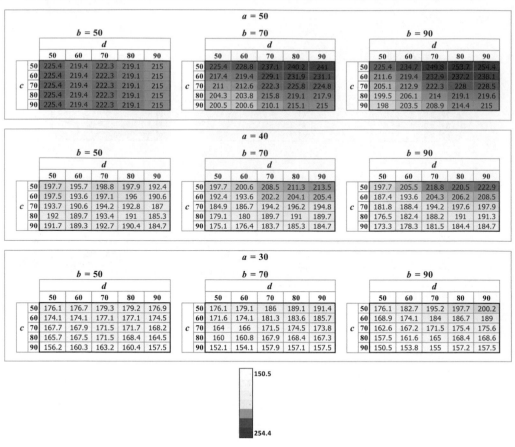

Figure A4 Average severity of drought paths obtained with different combinations of parameters. Severity is expressed as the ratio between the total sum of areas (in percentage) and duration (months).

a = 50

b = 50

c \ d	50	60	70	80	90
50	0	7	0	1	1
60	0	7	0	1	1
70	0	7	0	1	1
80	0	7	0	1	1
90	0	7	0	1	1

b = 70

c \ d	50	60	70	80	90
50	0	1	0	1	1
60	0	7	7	1	1
70	0	1	0	1	1
80	0	1	0	1	1
90	0	1	1	1	1

b = 90

c \ d	50	60	70	80	90
50	0	0	0	0	0
60	0	7	0	0	0
70	0	0	0	0	0
80	1	1	1	1	1
90	1	1	1	1	1

a = 40

b = 50

c \ d	50	60	70	80	90
50	1	1	1	1	1
60	7	7	7	1	1
70	0	7	1	1	1
80	7	7	7	1	1
90	7	7	7	1	1

b = 70

c \ d	50	60	70	80	90
50	1	1	1	1	1
60	7	7	7	7	1
70	0	1	1	1	1
80	7	7	7	1	1
90	7	7	7	7	1

b = 90

c \ d	50	60	70	80	90
50	1	1	4	0	0
60	7	7	7	7	7
70	0	1	1	1	1
80	7	7	7	1	1
90	7	7	1	1	1

a = 30

b = 50

c \ d	50	60	70	80	90
50	1	1	1	1	1
60	7	7	1	1	1
70	7	7	1	1	1
80	7	7	7	1	1
90	7	7	7	1	1

b = 70

c \ d	50	60	70	80	90
50	1	1	1	1	1
60	7	7	7	7	1
70	7	1	1	1	1
80	7	7	1	1	1
90	7	7	1	1	1

b = 90

c \ d	50	60	70	80	90
50	1	1	1	1	1
60	7	7	7	7	7
70	0	1	1	1	1
80	7	1	1	1	1
90	7	1	1	1	1

Figure A5 Mode of onset location of drought paths obtained with different combinations of parameters. Locations: centre (0), east (1), northwest (4), and south (7).

127

a = 50

b = 50

c	d 50	60	70	80	90
50	0	0	1	7	7
60	0	0	1	7	7
70	0	0	1	7	7
80	0	0	1	7	7
90	0	0	1	7	7

b = 70

c	d 50	60	70	80	90
50	0	0	0	0	0
60	0	0	0	0	0
70	1	1	1	7	7
80	7	7	7	7	7
90	3	1	7	7	7

b = 90

c	d 50	60	70	80	90
50	0	0	0	0	0
60	1	0	0	0	0
70	1	1	1	1	1
80	7	7	7	7	7
90	7	7	7	7	7

a = 40

b = 50

c	d 50	60	70	80	90
50	1	1	1	7	7
60	7	1	7	7	7
70	1	1	7	7	7
80	1	1	7	7	7
90	7	7	7	7	7

b = 70

c	d 50	60	70	80	90
50	1	1	1	0	0
60	7	1	7	7	7
70	7	1	7	7	7
80	7	7	7	7	7
90	7	7	7	7	7

b = 90

c	d 50	60	70	80	90
50	1	1	0	0	0
60	1	1	0	0	0
70	1	1	7	7	7
80	7	7	7	7	7
90	7	7	7	7	7

a = 30

b = 50

c	d 50	60	70	80	90
50	1	1	1	7	7
60	7	1	7	7	7
70	7	1	1	7	7
80	1	1	1	7	7
90	1	1	1	7	7

b = 70

c	d 50	60	70	80	90
50	1	1	1	7	0
60	7	1	7	7	7
70	7	1	1	7	7
80	7	7	7	7	7
90	1	1	1	7	7

b = 90

c	d 50	60	70	80	90
50	1	1	0	0	0
60	1	1	0	0	0
70	1	1	1	1	1
80	7	7	7	7	7
90	1	1	7	7	7

Figure A6 Mode of end location of drought paths obtained with different combinations of parameters. Locations: centre (0), east (1), north (3), and south (7).

128

a = 50

b = 50

c \ d	50	60	70	80	90
50	CCW	CW	CW	CW	CW
60	CCW	CW	CW	CW	CW
70	CCW	CW	CW	CW	CW
80	CCW	CW	CW	CW	CW
90	CCW	CW	CW	CW	CW

b = 70

c \ d	50	60	70	80	90
50	CCW	CW	CCW	CCW	CCW
60	CCW	CW	CW	CCW	CW
70	CW	CW	CW	CCW	CW
80	CW	CW	CW	CW	CW
90	CW	CW	CW	CW	CW

b = 90

c \ d	50	60	70	80	90
50	CCW	CW	CW	CW	CW
60	CCW	CW	CW	CW	CW
70	CW	CW	CW	CW	CW
80	CW	CW	CW	CW	CW
90	CW	CW	CW	CW	CW

a = 40

b = 50

c \ d	50	60	70	80	90
50	CCW	CW	CW	CW	CW
60	CCW	CW	CW	CW	CW
70	CCW	CW	CW	CW	CW
80	CCW	CW	CW	CW	CW
90	CCW	CW	CW	CW	CW

b = 70

c \ d	50	60	70	80	90
50	CCW	CW	CW	CCW	CW
60	CCW	CW	CW	CCW	CW
70	CCW	CW	CW	CW	CW
80	CW	CW	CW	CW	CW
90	CW	CW	CW	CW	CW

b = 90

c \ d	50	60	70	80	90
50	CCW	CW	CW	CW	CW
60	CW	CW	CW	CW	CW
70	CW	CW	CW	CW	CW
80	CW	CW	CW	CW	CW
90	CW	CW	CW	CW	CW

a = 30

b = 50

c \ d	50	60	70	80	90
50	CW	CW	CW	CW	CW
60	CW	CW	CW	CW	CW
70	CCW	CW	CW	CW	CW
80	CCW	CW	CW	CW	CW
90	CCW	CW	CCW	CW	CW

b = 70

c \ d	50	60	70	80	90
50	CW	CW	CW	CCW	CW
60	CW	CW	CW	CW	CW
70	CCW	CW	CW	CCW	CW
80	CW	CW	CW	CW	CW
90	CW	CW	CCW	CW	CW

b = 90

c \ d	50	60	70	80	90
50	CW	CW	CW	CW	CW
60	CW	CW	CW	CW	CW
70	CW	CW	CW	CW	CW
80	CW	CW	CW	CW	CW
90	CW	CW	CW	CW	CW

Figure A7 Mode of rotation of drought paths obtained with different combinations of parameters. Rotation is indicated by ccw and cw that sand for mostly counter-clockwise, and mostly clockwise, respectively.

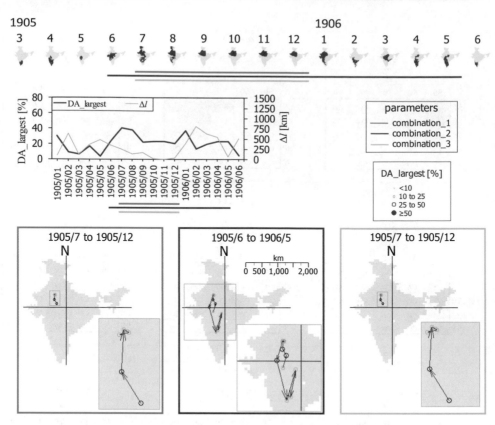

Figure A8 Results of drought of 1905. Largest clusters and centroids are indicated from 1905/3 to 1906/6 (top). Area of largest cluster (DA_largest) and distance between consecutive clusters in time (Δ*l*) are displayed from 1905/1 to 1906/6 (centre). The drought duration is pointed out schematically with a horizontal line for each combination of parameters. Drought tracks calculated with the three combination of parameters are also presented (bottom). Spatial drought extent is schematized by four symbols pointing out the size of area. The origin of the axes is placed in the centre of the country. Narrows point out the direction of each track segment. Insets show zoomed-in views.

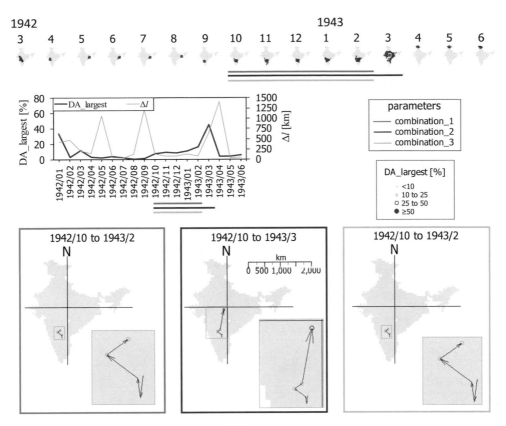

Figure A9 Same as Fig. A8 but for drought of 1942.

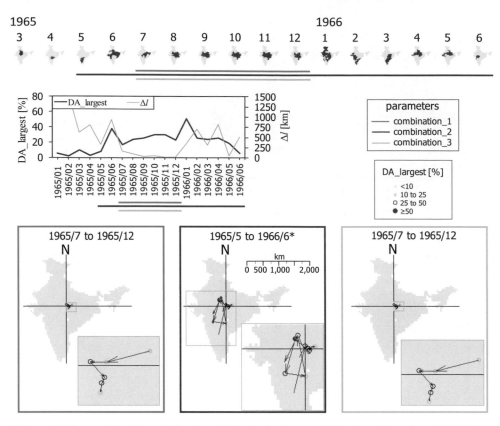

Figure A10 Same as Fig. A8 but for drought of 1965. *In the figure only it is shown the tracks until 1966/6 but they end in 1967/2.

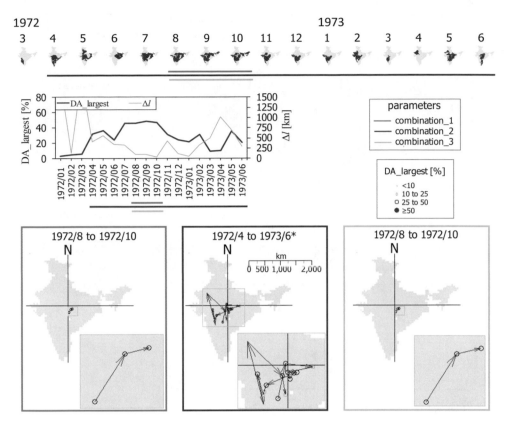

Figure A11 Same as Fig. A8 but for drought of 1972. *In the figure only it is shown the tracks until 1973/6 but they end in 1973/7.

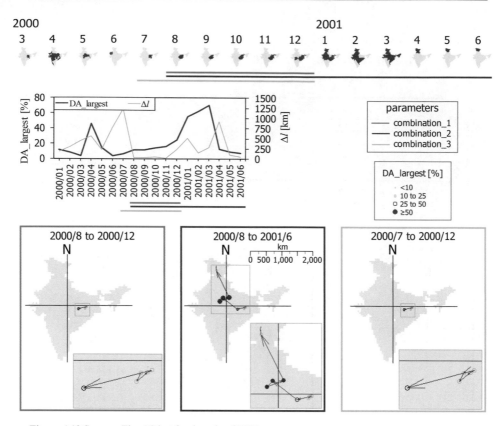

Figure A12 Same as Fig. A8 but for drought of 2000.

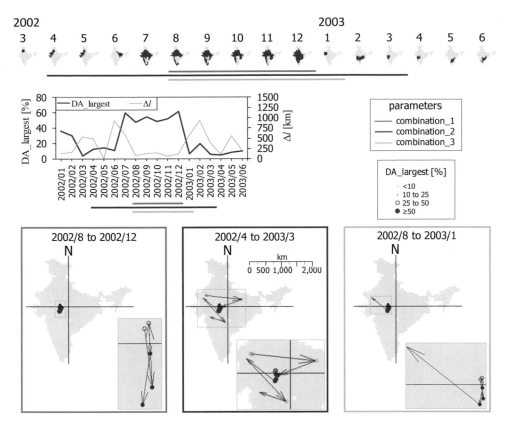

Figure A13 Same as Fig. A8 but for drought of 2002.

ABOUT THE AUTHOR

Vitali Díaz Mercado (Vitali Diaz) is a civil engineer, passionate programmer, data analyst, modeler, and remote-sensing-based approaches developer to overcome water challenges. Vitali is originally from Mexico. He holds a BSc degree in Civil Engineering and an MSc degree in Water Science from the Faculty of Engineering at the Autonomous University of Mexico State. He received his PhD from IHE Delft and the Delft University of Technology. His BSc thesis, MSc and PhD studies were financed and supported by the National Council for Science and Technology of Mexico.

He has collaborated on various projects with case studies in Mexico, Colombia, Ecuador, Dominican Republic, El Salvador, Honduras, Costa Rica, Mauritania, Senegal, Mali, Côte d'Ivoire, Burkina Faso, Tanzania, Greece, India and Vietnam. The Albert II of Monaco Foundation supported the last stage of his PhD through the project "Uncertainty-aware intervention design for Mediterranean aquifer recharge".

His research interests include extreme hydrological events (drought and flood), machine learning, data visualization, data analysis, data management, hydrological modeling, integration of models and remote sensing data, development of GIS-based applications and water accounting. These lines of research have arisen during different stages of Vitali's academic and professional journey.

Personal email: vitalidime@gmail.com

ORCiD: https://orcid.org/0000-0002-5502-4099

List of publications related to the doctoral research

Journals publications

Diaz, V., Corzo Perez, G. A., Van Lanen, H. A. J., Solomatine, D., and Varouchakis, E. A. (2020). An approach to characterise spatio-temporal drought dynamics. Advances in Water Resources, 137, 103512. https://doi.org/10.1016/j.advwatres.2020.103512

Diaz, V., Corzo Perez, G. A., Van Lanen, H. A. J., Solomatine, D., and Varouchakis, E. A. (2020). Characterisation of the dynamics of past droughts. Science of The Total Environment, 134588. https://doi.org/10.1016/j.scitotenv.2019.134588

Corzo, PGA, **Diaz, V.**, Laverde, M. (2018). Spatiotemporal hydrological analysis. International Journal of Hydrology, 2(1):25-26. doi: 10.15406/ijh.2018.02.00045

Book chapters

Diaz, V., Corzo, G., and Perez, J. R. (2019). 3 - Large-scale exploratory analysis of the spatiotemporal distribution of climate projections: applying the STRIVIng toolbox. In G. Corzo and E. A. Varouchakis (Eds.), Spatiotemporal Analysis of Extreme Hydrological Events (pp. 59–76). Elsevier. https://doi.org/10.1016/B978-0-12-811689-0.00003-3

Diaz, V., Corzo, G., Lanen, H. A. J. Van, and Solomatine, D. P. (2019). 4 - Spatiotemporal drought analysis at country scale through the application of the STAND toolbox. In G. Corzo and E. A. Varouchakis (Eds.), Spatiotemporal Analysis of Extreme Hydrological Events (pp. 77–93). Elsevier. https://doi.org/10.1016/B978-0-12-811689-0.00004-5

Le, H. M., Corzo, G., Medina, V., **Diaz, V.**, Nguyen, B. L., and Solomatine, D. P. (2019). 7 - A comparison of spatial–temporal scale between multiscalar drought indices in the South Central Region of Vietnam. In G. Corzo and E. A. Varouchakis (Eds.), Spatiotemporal Analysis of Extreme Hydrological Events (pp. 143–169). Elsevier. https://doi.org/10.1016/B978-0-12-811689-0.00007-0

Conference papers

Diaz, V., Corzo Perez, G. A., Van Lanen, H. A. J., and Solomatine, D. (2018). Intelligent drought tracking for its use in Machine Learning: implementation and first results. (G. La Loggia, G. Freni, V. Puleo, and M. De Marchis, Eds.), HIC 2018. 13th International Conference on Hydroinformatics (Vol. 3). Palermo: EasyChair. https://doi.org/10.29007/klgg

Diaz Mercado, V., Corzo Perez, G., Solomatine, D., and Van Lanen, H. A. J. (2016). Spatio-temporal analysis of hydrological drought at catchment scale using a spatially-distributed hydrological model. Procedia Engineering, 154, 738–744. https://doi.org/10.1016/j.proeng.2016.07.577

Conference abstracts

Diaz, V., Corzo Perez, G. A., Van Lanen, H. A. J., and Solomatine, D. (2018). Comparative analysis of two evaporation-based drought indicators for large-scale drought monitoring. Geophysical Research Abstracts Vol. 20, EGU2018-18728. EGU General Assembly. Vienna.

Osman, A., **Diaz, V.**, Corzo Perez, G. A., Varouchakis, E., Solomatine, D. (2018). Finding negative response of crop yield to drought: a spatiotemporal approach over East India. International Conference on Water, Environment, Energy and Society (ICWEES), Tunisia. Based on Ahmed's MSc Thesis

Diaz, V., Corzo G., Van Lanen H.A.J., Solomatine D. (2017). On the visualization of water-related big data: extracting insights from drought proxies' datasets. Geophysical Research Abstracts Vol. 19, EGU2017-10718-1. EGU General Assembly, Vienna

Diaz, V., Corzo Perez G., Van Lanen H.A.J., Solomatine D. (2016). Spatio-temporal analysis of large-scale meteorological drought: helping to achieve the SDGs 6.A and 11.5. 12th Kovacs Colloquium, Paris, France. DOI: 10.13140/RG.2.1.2595.2888

Co-supervisor of MSc research

Spatiotemporal analysis and prediction of crop yield using data-driven models and drought areas. Case study of India. Ahmed Abdelmoneim Ahmed Osman. MSc Thesis. WSE-HERBD.18-17, IHE-Delft, March 2018. Delft, Netherlands

Integrated spatial precipitation drought index by combining remotely sensed information and local stations. Case study Guerrero State, Mexico. Yousra Omer Elfaroug Mohammed Khair. MSc Thesis. WSE-HI. 16-03, IHE-Delft, April 2016. Delft, Netherlands

Netherlands Research School for the
Socio-Economic and Natural Sciences of the Environment

D I P L O M A

for specialised PhD training

The Netherlands research school for the
Socio-Economic and Natural Sciences of the Environment
(SENSE) declares that

Vitali Díaz Mercado

born on 8 June 1983 in Toluca, México

has successfully fulfilled all requirements of the
educational PhD programme of SENSE.

Delft, 24th November 2021

Chair of the SENSE board   The SENSE Director

Prof. dr. Martin Wassen   Prof. Philipp Pattberg

The SENSE Research School has been accredited by the Royal Netherlands Academy of Arts and Sciences (KNAW)

KONINKLIJKE NEDERLANDSE
AKADEMIE VAN WETENSCHAPPEN

The SENSE Research School declares that Vitali Díaz Mercado has successfully fulfilled all requirements of the educational PhD programme of SENSE with a work load of 45.3 EC, including the following activities:

<u>SENSE PhD Courses</u>

o   Environmental research in context (2015)
o   Research in context activity: 'Co-organizing 36th IAHR World Congress on: Deltas of the future and what happens upstream (World Forum, The Hague – 28 Jun to 3 Jul 2015)'

<u>Selection of Other PhD and Advanced MSc Courses</u>

o   Model building, inference and hypothesis testing in hydrology, Luxembourg Institute of Science and Technology (2015)
o   Modelling theory and Computational Hydraulics, UNESCO-IHE (2015)
o   English for Academic Purposes, TU Delft (2015 and 2016)
o   Managing the Academic Publication Review Process, TU Delft (2016)
o   Popular Scientific Writing, TU Delft (2019)
o   Voice training, TU Delft (2019)
o   Presenting Scientific Research for PhD students, TU Delft (2019)

<u>Management and Didactic Skills Training</u>

o   Member of IHE-PhD fellows Association Board and the SENSE PhD council (2016-2017)
o   Co-organisation of the IHE PhD symposium 2017
o   Supervising two MSc student with thesis (2016-2018)
o   Teaching in the MSc course 'Geostatistics for Water Management and Environmental Sciences' (2018)
o   Teaching in the MSc course 'Data Management and Visualization Tools for Water Resources' (2019)

<u>Oral Presentations</u>

o   *A framework for regional and global long term and middle term assessment of floods and droughts*. 36th IAHR 2015 Congress, 28 June-3 July 2015, The Hague, The Netherlands
o   *Spatio-temporal analysis of hydrological drought at catchment scale using a spatially-distributed hydrological model*. 12th HIC 2016 Conference, 21-26 August 2016, Incheon, Korea
o   *On the visualization of water-related big data: extracting insights from drought proxies' datasets*. EGU General Assembly, 23-28 April 2017, Vienna, Austria
o   *Intelligent drought tracking for its use in Machine Learning: implementation and first results*. 13th HIC 2016 Conference, 1-6 July 2018, Palermo, Italy

SENSE coordinator PhD education

Dr. ir. Peter J. Vermeulen